i walk therefore I AM

How Walking on Two Legs Proves That God Exists

Jerry Sweafford, Jr.

i walk therefore I AM: How Walking on Two Legs Proves That God Exists

Jude and Niara:

May you never find yourselves at the fence.

Table of Contents

1 - It Matters That God Exists

Is it, you don't believe, or you won't believe?[1]

Jerry Flowers, Jr.

I have a young friend (let's call her Susannah) who regularly tells me about her adventures in Unicornland. This is a land of gravity-defying, flying horses and rainbow-colored fruity drinks that taste amazing. Often, I ask Susannah, "Did you go to Unicornland last week?" She then tells me about everything that has happened since I last asked. In a recent update, she made a face while telling me about a Pegasus she encountered.

Why the face? I wondered aloud. She gently explained, as if to a child much younger than herself, that a Pegasus isn't as great as the unicorns are. So, you see, Unicornland is an exciting and fascinating place, but there's just one problem:

Susannah is the only person that has ever been there.

---

[1]Your God Given Assignment [2024]: https://www.youtube.com/watch?v=L3z0kJ4QrKM&t=165s

The atheist views the believer's faith in God very similarly to the way most people think of Susannah's descriptions of Unicornland. He might even assert that there is MORE evidence of Susannah's adventures than there is of the existence of God.

Is that true?

Is God simply a figment of one's imagination?

A cosmic Santa Claus?

An emotional crutch?

NO.

But HOW can one explain an invisible God to a person who has reduced the probability that God exists to that of the Tooth Fairy, Easter Bunny, or, yes, a unicorn? Well, that brings me to why I wrote this book in the first place.

Why I Wrote This Book

Biomimicry is a design approach in which engineers find inspiration for their products from what they observe in nature. I work in the field of robotics and artificial intelligence (AI). I often come across evolutionary explanations of nature's systems that oppose basic logic, reason, and science. In my doctoral research work, I designed controllers that used AI to learn how to control a humanoid robot so that it would stand and walk in a balanced way.

Since humanoid robots are human-like, I used a biomimetic approach. Studying the human body helped me to understand principles of balance for developing joint controllers for the robot. I found research papers that would state (as a matter of fact) that humans came from ancient primates that walked on four limbs. Those papers never gave actual evidence for such a claim.

They did not provide a reasonable explanation for four-legged primates developing the ability to balance and walk on two legs. From their perspective, it happened, so there's no need to question it. It was annoying to see such a weak hypothesis given as a matter of fact, when there are more reasonable alternative explanations. That isn't science, or at least, it shouldn't be.

This topic is not directly connected to my research work. Yet, I wrestled with writing an appendix chapter in my dissertation about it. I finally decided to finish my PhD, and write an essay about it later. That essay morphed into this book.

Often, books on the existence of God overwhelm readers with many examples of design in the world, presumably to be more convincing. I suspect that a thorough explanation of a single example of design might be a more effective approach.

In this book, we will closely examine a single example of design in nature that you regularly experience or observe: humans walking on two legs. My thesis is simple: *the existence of God is the most reasonable explanation for the incredible ability of humans to balance and walk on two legs.*

Who This Book is For

I wrote this book for those who are at (what I call) the Fence. Imagine a Fence separating two properties. On one side, there is a barren, deserted area with dry, dead trees and no grass or greenery.

On the other side of the Fence, there is a cool, lush orchard full of colorful fruit and plants. A curious atheist stands at the Fence on the desert side, gazing over at the orchard. He wonders if he should step through to the other side, or walk back toward the desert land to which he is accustomed.

A few yards away, an uncertain believer has wandered away from the orchard and is at the Fence, inquisitively observing the desert land on the other side. She is unsure: should she go back toward the orchard, or is she missing out on something on the other side of the Fence?

Which Side of the Fence Are YOU On?

Uncertain Believers

It is common to encounter believers with serious doubts about their faith. And, it is increasingly likely to find believers who are not at all confident that God exists. One of the most alarming trends is that of college-age believers abandoning the faith of their teenage years.

Surveys consistently report that anywhere from 50% to 80% of young people stop attending church or leave the faith during their college years. Often, this occurs because they encounter the naturalistic worldview for the first time. I have personally asked middle and high school students in the church about concepts like evolution and materialism. I have generally found that they had little to no knowledge of these ideas.

This book is not a primer on materialism or evolution, but I will address major ideas from them. The naturalistic worldview opposes what we observe in nature and science. And, of course, it is incompatible with faith in God. We should educate young believers earlier on these kinds of ideas.

College-age believers can be blind-sided by this worldview when they first encounter it. They are often misled by lies and false assumptions. Many are persuaded to believe in the opinions of the scientific majority. I wrote this book to help doubting believers turn away from the Fence, and return to the land of faith.

Curious Atheists

Professor Sy Garte, who holds a PhD in biochemistry, grew up as a third-generation atheist in a politically left-wing family that embraced Communism. He admits that as a young man in college, he felt that something was missing. He used science to fill that void and give him some sense of purpose:

But there was one catch, which was that what I was learning in science didn't jibe with the materialistic worldview... For example, in graduate school, I learned about the process by which proteins are made in cells, and that's a very complex process that involves a tremendous amount of biomolecules interacting with each other and the complexity is just incredible. And, I remember feeling like a chill going down my spine, like...this is amazing! When did... how did this get here? And it was something that I couldn't answer. Later, of course, like most biologists, I just came up with the answer, which is evolution, which does everything...[2]

[2]Why This Atheist Scientist Became a Believing Christian [2022]: https://www.youtube.com/watch?v=OMBQwGzn_TE&list=PLGZlwR-miLjm3p7xYJN9Z9CoPA-MwTQm4n&t=190s

Many agnostics and atheists are similarly amazed when observing the world. They have also been compelled to believe that faith is incompatible with "capital-S" Science. They know that there is something or Someone beyond "just this." But, the culture pushes them away from the possibility of faith.

This book offers a simple but powerful proof of the existence of God. The curious atheist will be able to step forward into the land of faith. Unexpectedly, he will find that his intellect and reason will be welcomed there, not left behind in the realm of unbelief.

For My Atheist Friends

After interviewing several former believers, a researcher found a repeating pattern. Her finding was that there are five phases that lead to the rejection of one's faith: Detachment, Doubt, Dissociation, Transition, and Declaration[3]. I realized that the atheist's journey from unbelief to faith follows the same pattern, but in reverse. One observes the same phases from skepticism to theism:

[3]Julie Krueger, *The Road to Disbelief: A Study of the Atheist De-Conversion Process* [2013]

- Detachment: you sense a vague dissatisfaction or discomfort with unbelief.

- Doubt: you have identified specific areas of discomfort or dissatisfaction with atheism.

- Dissociation: you have rejected the atheistic identity.

- Transition: you are testing out theism.

- Declaration: you have accepted that a Higher Power exists.

If you are a curious atheist standing at the Fence described earlier, then you are either detached from, or doubtful of, atheism. You are hesitant about continuing forward. You think to yourself: **how can I explain this journey to faith to my atheist friends without sounding brainless or simple-minded?**

This book will help you through the final stages leading to belief. You will have a simple yet rational way to show why your thinking has changed. You will be able to show your friends that the existence of a Creator is the only rational explanation for how humans walk on two legs. Then, you can ask them to give you a better explanation for how humans balance and walk like we do. When they are unable to do so, you will feel much better.

How to Read This Book

Skeptics who are worried about wasting their time should take what I call the *"don't or won't"* test. If you cannot think of any reasonable evidence that could convince you of the existence of God, then you have decided not to believe. In other words, you won't believe, and reading this book will be irritating and frustrating for you. Otherwise, you simply don't believe, and will be open to a fresh perspective.

Part 1 (Chapters 2-4) makes the case for a Designer of the human body. In chapter 2, we'll learn how difficult it is to walk on two legs. We'll explore how humans have optimized bodies to do so, and that we have an amazing ability to learn to walk from scratch. In chapter 3 we examine the popular evolutionary story of an ancient quadruped turning into the modern-day human. We'll answer the question: did that actually happen?

In chapter 4, we will apply the design inference to the human body's ability to balance and walk on two legs. We'll specify attributes of human bipedal walking that point to a Designer. Then, we'll show that scientists are becoming more vocal with theistic hypotheses than they were in the late 20th century.

Part 2 (Chapters 5-9) is a resource, especially for the agnostic, the atheist, and the doubting believer. Eager believers can present their faith in a way that leaves gaps of understanding for the person who disagrees. It's partly why I named the book the way I did. We believers can often say things like, "Look at the design of the universe. Therefore, God!" or "The irreducible complexity of this organelle of the cell proves God!" Well, not quite.

The atheist, agnostic, and doubter generally don't agree with the believer's initial assumptions. So, they will not reach the same conclusion as the person of faith. I fill these gaps in part 2. In chapters 5 and 6, we establish clear definitions of God and faith. In chapters 7 and 8, we set a standard for reasonable evidence of God's existence. We also define what it means to make a valid proof of God's existence.

I will challenge you to set standards of evidence and proof before continuing in the book. This makes sense because you will likely disagree with my conclusions if we cannot agree upon these standards first. Finally, I make the case for Yahweh as the Creator of two-legged humans in chapter 9. In that chapter, we also eliminate other deities and the universe as potential sources of the universe.

In chapter 10, I end by anticipating a question from now-former agnostics or doubters: Now What? We'll show that trust in God is far more than belief in a fact. The person of faith decides to base his or her entire life on this Truth. The book ends with a discussion of the personal significance of faith in God.

2 - Walking on Two Legs is Really Hard to Do

[I]magine trying to write a computer program to make a robot walk–again, a fiendish problem.[4]

Shane O'Mara

For my doctoral research, I designed balance and gait controllers for simulated and actual humanoid robots. My research focus was what is known as push recovery. As the term suggests, push recovery deals with moving in a way to avoid falling after being pushed. We humans do well with push recovery. I can push you unexpectedly (not too hard) and you will reflexively adjust your posture and take a step, if necessary.

People often nod in agreement when I tell them this. They are somewhat surprised about what I say next: you have no idea how you will perform the push recovery motion. You cannot predict what you will do. So, we can have 100% confidence that you won't fall if I push you. But, there is no way to predict what you will do to recover from my push.

---

[4] In Praise of Walking: A New Scientific Exploration, 2020.

We balance and walk on two legs very well, but we don't know exactly what we do to maintain balance while standing and walking. That was a problem for me. I wanted to study bipedal balance in humans so that I could program humanoids to do what we do so well. I focused on the software side of getting robots to balance.

As I studied human balance, I found several anatomical features that allow humans to efficiently and robustly walk on two legs. A major part of what makes us so good at balancing on two legs is that our bodies have an optimal structure for bipedal gait. Yet, walking isn't just a way for us to move around. It is also an essential part of good health.

Furthermore, walking has more benefits for humans than exercise alone. Walking calms me, helps me think through problems, and just makes me feel better. I am not the only person that feels this way. Going too long without a walk, and I just feel like... bleh.

Evolutionary biologists and paleoanthropologists supply a popular explanation for human bipedalism.

The evolutionary story is that quadrupedal primates evolved into the bipedal primates we now call humans. The associated rewards of two-legged balance and walking are happy coincidences.

There is a much older, alternative perspective that has been rebranded in recent years. From that perspective, human bipedalism and its advantages are evidence of purposeful design.

Which is it?

In the next three chapters we will answer that question. In this chapter, we examine three important components of bipedal balance in humans.

In chapter 3, we will discover three problems with the popular evolutionary story. We'll see that it doesn't adequately explain the amazing walking machine known as the human body. Then, in chapter 4, we will focus on two aspects of human balance that point directly to a creative Designer.

There are three things that make reliable walking on two legs such a remarkable ability:

1. Balancing on two (rather than three or four) legs is unstable.

2. The human body is structured for making two-legged balance easier.

3. The nervous system of every healthy human learns to balance its body on two legs.

Difficult Dynamics

You've probably seen or heard of a three-legged stool or a table with at least four points of contact with the floor. But, have you ever heard of a two-legged chair or desk?

Why not?

In the following image, the four-legged stool on the left is clearly more likely to stay upright than the two-legged stool on the right.

The reason we make stools and tables with at least three legs is that these are stable configurations. By "stable", we simply mean that having three (or more) contact points will allow a desk or chair to stay upright with no external effort.

We see the same thing with vehicles. Some small children start with riding a tricycle, because of its stability. I say "some kids", because I'm not sure that my nephew rides a tricycle. He has a little battery-powered four-wheeler that he rides around his neighborhood. He thinks it's hilarious when Tio (me: "Tio" is the Spanish word for "uncle") climbs in and begins to ride. It slows down considerably, but I am impressed that his little four-wheeler can carry me uphill!

Anyway, the kids who ride tricycles later transition to a bicycle with training wheels and finally learn to ride a bike with no training wheels. At each stage, the decreasing stability of the vehicle must be overcome by increasing the skill of the rider. Tricycle riders don't have to think about balancing at all, but bike riders do. Try to stand a bike up without the kickstand and it will immediately fall.

To use the language of physics and engineering, the tricycle is a stable vehicle, but a bicycle is inherently unstable. This discussion of two-wheeled vehicles and two-legged furniture also applies to living things. Four-legged animals stabilize themselves more easily than two-legged animals do. Having more contact points with the ground creates a more stable arrangement.

It's why babies first learn to crawl, which gives them at least four contacts with the ground. Then, they "cruise", which means to walk while also making contact with a wall, couch, or some other surface. Finally, they learn to walk unassisted. A decrease in stability is why many elderly people transition from unassisted walking to using a cane or walker. The additional contact points give them better stability.

Healthy children and adults (with no disability or injury) balance themselves and walk on two legs without assistance. But how can humans be so good at such an inherently unstable behavior? There are two ways. We each have a body that makes it easier to balance, and we have a nervous system that controls that body very well. Let's consider the body first.

Specialized Hardware

The human body makes it easier to balance on two legs. To show this, it helps to look at a few key gait-related anatomical differences between quadrupedal primates and humans. Table 1 (on page 27) presents several features in humans that make for efficient bipedalism[567]. Notice that none of these human features that aid in two-legged walking exist in any of the primates that are most like humans.

For humans, the foramen magnum (the opening at which the brain connects to the spinal cord) is at the base of the skull. This positions the head above the spine and is helpful for upright standing and walking. Humans have a barrel-shaped rib cage. This allows their shorter arms to freely swing to the side. Quadrupedal primates have longer arms and cone-shaped rib cages that allow for overhead reaching.

Quadrupedal primates have a rear-skull spinal attachment, long, narrow pelvis, and short, bent legs. This gives them a more horizontal inclination. Humans have a broad, short pelvis and long, extendable legs for bipedal gait.

---

[5]Foley and Lewin. Principles of Human Evolution, 2004.

[6]Roger Lewin. Human Evolution: An Illustrated Introduction, 2005.

[7]The Complete Human Body: The Definitive Visual Guide, Alice Roberts, ed., 2016.

Table 1: Gait-related anatomical differences between quadrupedal primates and humans

Quadrupedal primates	Feature	Humans
Rear of skull	Spine attachment	Bottom of skull
Long, overhead-reaching	Arms	Short, side-swinging
Cone-shaped	Rib cage	Barrel-shaped
Narrow, long	Pelvis	Broad, short
Bent, short	Legs	Extendable, long
Grasping	Foot	Acts as lever
Opposable, smaller	Big toe	Adducted, larger

It is not generally believed that chimpanzees are ancestors of humans. But, the evolutionary story is that chimps and bipedal humans share a close, common ancestor. Three key feature differences between humans and chimpanzees contribute to their gait differences.

1. Chimpanzees can't extend the knee to make a straight leg. Extendable legs give humans a more efficient gait. Because of this, we need

less muscular power to support the body with a straight leg during the stance phase.

2. Chimpanzees have femurs that do not slope inward, so their feet are farther apart. This causes a waddling gait. Humans walk with far less lateral shifting of the center of gravity.

3. Chimpanzees have grasping feet, but humans have transverse and longitudinal foot arches. These allow the foot to act as a lever that can propel the body forward.

There are significant differences between chimpanzees and humans. This suggests that a common ancestor would also be much different than modern humans as shown in the following figure.

Quadrupedal Ape vs. Bipedal Human

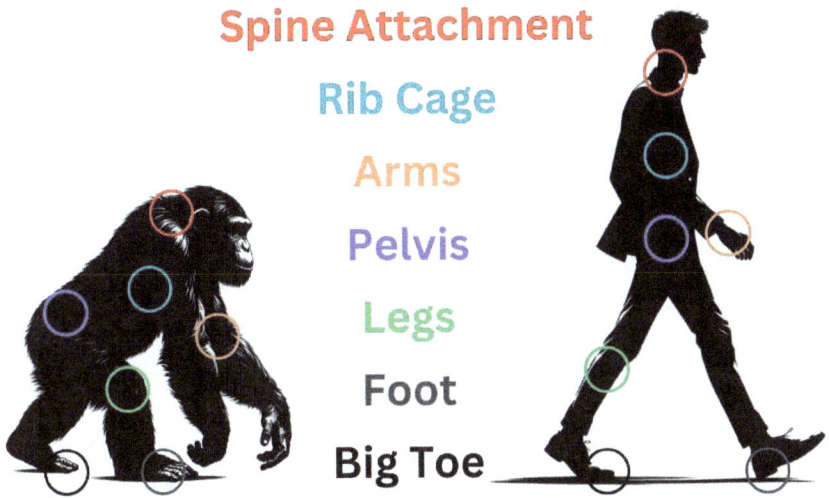

Spine Attachment

Rib Cage

Arms

Pelvis

Legs

Foot

Big Toe

Integrated Software

For me (as a robotics engineer and AI researcher), the "software" aspect of human balance and gait is the most incredible part. Unicycles, bicycles, and motorcycles are interesting vehicles, but what really fascinates us is when an expert rider controls these vehicles skillfully.

It is interesting that the human body has all of the optimized hardware to make it easier to stand and walk on two legs, but without the right internal controller (our nervous system), the body would still fall down. Because the bipedal arrangement is inherently unstable and there are so many ways to fall down, the fact that we rarely fall is amazing! Think about it: when was the last time you fell down? You probably can't remember.

But, how often have you slipped, tripped, or bumped into something or someone in the last week? You do it all the time. But you don't fall. As someone who has studied humanoid push recovery for over a decade, I can tell you that we are nowhere near that level of robustness with our gait and balance controllers in bipedal robots.

Back in 2015, the DARPA Robotics Challenge Finals was the culmination of a three-year international competition designed to stimulate innovation in robotics for disaster scenarios. But, instead of showing the world the exciting potential of robots, the competition is perhaps best known for humorous videos of bipedal robots falling down[8]. To be clear, the robots in that event were incredible. The issue

---

[8] A Compilation of Robots Falling Down at the DARPA Robotics Challenge [2015]: https://www.youtube.com/watch?v=g0TaYhjpOfo

was with controlling them.

It is a very difficult problem to control a robot in different environments and terrains, and having it touch or bump into things without falling down. Even if we solve this problem next week, that won't negate the fact that it is a very hard problem. But, it's a problem that we humans rarely think about. We navigate the world effortlessly on two legs.

It's also interesting that gait is such a unique characteristic, that we can differentiate between people using gait, similar to biometric identification using fingerprints. Yet, although every gait is unique, we all are able to walk and balance effectively.

There are quite a few four-legged animals that begin to walk on the same day on which they are born. Unlike horses, pigs, or other animals in that category, humans do not have the physical or neurological ability to walk until many months after birth.

In my doctoral research work, I came up with ways to train artificial neural networks to "learn" balance for simulated humanoids from scratch. This means that the robot always falls at first, but eventually learns either to stand for longer and longer periods or to walk for farther and farther distances. The problem was that there was really no way to ensure that the neural networks would converge to a good solution.

In fact, many times they didn't. This makes what happens in us humans so amazing. To repeat from earlier, all healthy children and adults without disability or injury (by definition) balance themselves and walk on two legs without assistance. And we do this *from scratch*, meaning that everyone learns to do it, and no one can do it just after they are born, like horses can.

I know you walk around all the time, and see everyone else doing it, so it might make sense to disregard walking and balancing on two legs as an easy thing. I want to assure you that it is not easy, and even suggest that you should really view how we walk the same way you would think of a unicyclist on a tightrope. With amazement. What we do when we walk and balance on two legs is skillful.

In this chapter, I wanted to show three things. First, balancing and walking on two legs are difficult to do. Second, the human body has special "hardware" that makes two-legged balance easier. Third, humans learn how to uniquely control our special bodies to overcome the instability of two-legged balance.

How do we explain this? In the next chapter, we will examine the popular explanation of neo-Darwinian macroevolution. We will show well it fits with what we have just learned about humans on two legs.

CHAPTER 2 SUMMARY

- The two most common explanations for the design of the human body are neo-Darwinian macroevolution and intelligent design.

- It is really hard to balance and walk on two legs. It is much easier to fall down than it is to balance well.

- Humans have specialized bodies for bipedalism, compared to quadrupedal primates.

- Humans have integrated nervous systems that allow us to effectively learn to use our specialized "hardware" to balance and walk on two legs.

How do we best explain the amazing ability of humans to balance and walk on two legs?

3 - Walking on Two Legs is Not the Product of Evolution

To suppose that the eye, with all its inimitable contrivances for adjusting the focus to different distances, for admitting different amounts of light, and for the correction of spherical and chromatic aberration, could have been formed by natural selection, seems, I freely confess, absurd in the highest possible degree.[9]

Charles Darwin

Have you read Darwin's <u>On the Origin of Species</u>?

I have, and it gave me a greater appreciation for Darwin. He was trying to correct a common but wrong belief that was prevalent during his time. Many people believed that each species in existence had been independently created. Darwin showed that all species within a genus shared common ancestry. He also showed that natural selection was the primary way that species change over time. **No one today disagrees with that idea.**

---

[9]<u>On the Origin of Species By Means of Natural Selection, or the Preservation of Favoured Races in the Struggle for Life</u>, 1859.

I repeat: nobody (including believers) disputes the fact that there are small changes among species. So, the big debate is not over whether evolution occurs. Of course it does! The mistake Darwin made was in extrapolating evolution in ways that he had no evidence for.

Darwin knew he was presenting ideas that would cause trouble, but he wasn't arrogant about it. As the quote opening this chapter suggests, Darwin had humility. He stated that his theory seemed absurd when applied to actual biological mechanisms such as an eye.

The general theory of evolution by natural selection has six main components: evolution, gradualism, speciation, common ancestry, natural selection, and nonselective mechanisms of evolutionary change[10]. Most of these concepts are universally accepted. Two of them, speciation and common ancestry, are not.

Gradual changes in species over time by natural selection and nonselective means are observable. But,

---

[10]Jerry A. Coyne, Why Evolution is True, 2009.

the ideas of common ancestry and speciation are an extrapolation of the existing evidence. Extrapolation means to extend a projection (like a trendline or curve) to an unknown range or situation for which there is no data. For example, let's say that I have lost thirty pounds over the past month.

An extrapolation might be to go forward in time five months and assume that I will be 150 pounds lighter than I am now! As the example chart on the next page shows, the trend of the past four weeks is not necessarily the trend of the next five months or the previous twelve months. If its underlying assumptions can be verified, then "extrapolated evolution"[11] is a valid hypothesis. Otherwise, it should be rejected, as many extrapolations are.

---

[11]Sometimes, this is called macroevolution (which literally means, "big" evolution) to distinguish it from the microevolution ("small" evolution) we actually observe throughout nature.

Extrapolation

WHY YOU SHOULDN'T DO IT

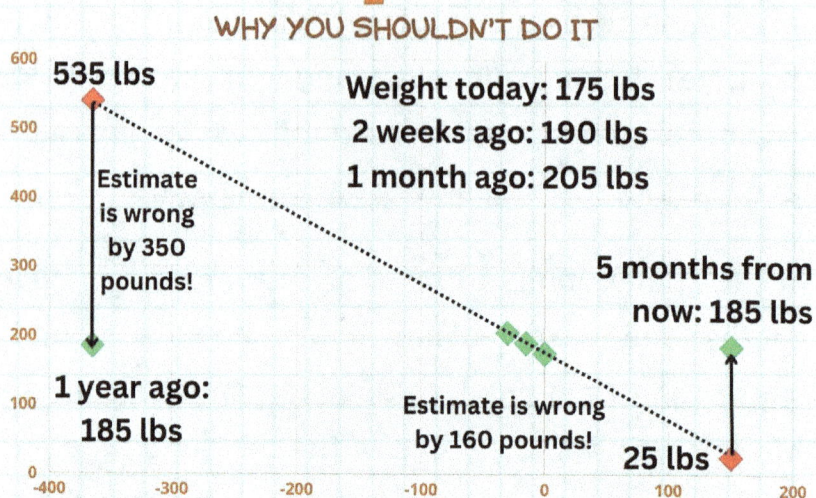

Perhaps the most fundamental evolutionary assumption is that similar traits in living species (such as DNA) suggest that they share a common ancestor. But, this assumption begs the question. To assume that the evolutionary trend goes back to a common ancestor for all species does not come from the evidence. You actually have to assume common ancestry first and then interpret the evidence that way.

Common traits among living things could actually suggest a common Designer! But, that's for the

38

next chapter. In this chapter, we will not assume that speciation and the common ancestry of all living things are true.

Instead, we will ask, how well does the evolutionary story explain how humans balance and walk on two legs? We will see that evolution could not have produced bipedal balance in humans. Then, we will show that scientists are increasingly doubtful of neo-Darwinian macroevolutionary hypotheses.

Evolution Could Not Have Produced Bipedal Balance in Humans

The evolutionary explanation for human obligate bipedalism suffers from three flaws.

1. The fossil record does not contain a high number of transitional forms. The fossil record should show many quadrupeds that were capable of two-legged walking. We would observe this, if macroevolution occurred. Yet, the large number of "missing links" does not exist.

2. The sequence of evolutionary changes needs a matching series of selective pressures. But, it is common to offer only one pressure as the driver behind the adaptations. Furthermore, there is no consensus on this point. There may be as many explanations for two-legged walking as there are researchers to propose them!

3. The proposed mechanism for the needed changes could not have produced them.

Lack of Transitional Forms is Indefensible

If humans came from four-legged ancestors, there should be lots of evidence in fossils. Darwin thought we would find fossils of in-between forms, both living and extinct. When he noticed these forms were missing, he gave three reasons. First, he said nature might not have preserved these "missing links." Second, he suggested scientists might have misclassified them. Third, he warned against making conclusions based on gaps in the fossil record[12].

---

[12]On the Origin of Species By Means of Natural Selection, or the Preservation of Favoured Races in the Struggle for Life, 1859.

Some people think Java Man is a missing link between apes and humans. Java Man was discovered in 1891 and included a tooth, a skull fragment, and a leg bone. The leg bone might suggest walking on two legs, and the skull size is between that of a human and a chimpanzee[13]. Eugene Dubois, who found these bones, called them *Pithecanthropus erectus*. The modern term is *Homo erectus*. Dubois's discovery shows fossil interpretation can be subjective.

One researcher might find a few bones and say they belong to a new species. Another researcher might see them differently. One professor pointed out that if today's dog breeds were only known as fossils, scientists might classify them as many different species, not just one[14].

What Darwin claimed about misclassified fossils applies to those classified as pre-human ancestors, such as australopithecines like "Lucy". Fossils that

---

[13]Jeremy DeSilva, First Steps: How Upright Walking Made Us Human, 2021.

[14]Jerry A. Coyne, Why Evolution Is True, 2009.

appear to be transitional might come from abnormal humans or apes.

The idea of missing fossils continues today. Evolutionists say there is a common ancestor for any two species. But, one professor wrote that there is almost zero chance of finding that ancestor in fossils[15]. One must believe in common ancestry, even without direct evidence.

If human evolution happened as recently as suggested, fossils should show a common ancestor of chimpanzees and humans. There should, at least, be a clear path from that ancestor to modern humans. Instead, scientists have found only a few fossils that can definitely be called ancient in-between forms. And, "we are left only with rare, fragmented bones to reconstruct the life of our ancestors"[16].

Darwin warned against making conclusions based on gaps in fossils. To be consistent, we shouldn't accept the idea of a single common ancestor until there is fossil evidence for it. It's possible, but not

---

[15]Jerry A. Coyne, Why Evolution Is True, 2009.

[16]Jeremy DeSilva, First Steps: How Upright Walking Made Us Human, 2021.

likely, that the fossil record will show the many in-between forms predicted by evolution. While we wait, it's useful to think about why these forms might have existed, especially regarding how humans developed two-legged walking.

Proposed Selective Pressures are Insufficient

Reasons for humanity's development of bipedalism abound, both in academia and popular culture. Few hypotheses on the development of bipedalism in humans are taken seriously. Many, if not most, of them are not testable. For example, Richard Dawkins famously suggested that bipedalism spread among primates because it was fashionable[17]. That hypothesis is not testable and has not been widely accepted.

Three hypotheses about the origins of human bipedalism are very popular.

1. The first highlights the benefits of freeing the hands. This could help with carrying children, tools, or weapons.

---

[17]Richard Dawkins (@RichardDawkins), X (then Twitter) post, March 2, 2019.

2. The second focuses on upright standing. This might help with staying cooler or getting a better view of one's surroundings.

3. The third suggests that two-legged walking is more energy-efficient than four-legged walking. This might be helpful in activities such as tracking prey over long distances.

But, these popular explanations only explain the *eventual* adoption of two-legged walking. They lack a *progression* of pressures to force quadrupeds toward two-legged walking.

In rare cases, researchers offer a hybrid hypothesis. For example, in Hunt's synthesis of hypotheses, bipedalism began as a ground-based feeding posture. This freed the hands to reach into trees. Later, it evolved into locomotion for more energy-efficient, long-distance travel between feeding sites[18]. But the hybrid hypotheses do not rationalize pre-human ancestors discarding four-legged walking.

---

[18]Kevin D Hunt, *The evolution of human bipedality: ecology and functional morphology*, 1994.

Unless they were responding to a *series* of pressures, these primates would not have stopped walking on four limbs. Their bodies were well adapted to it. To discard it would have made them less suited to their environment. Their bodies were not optimized for bipedal locomotion. Adopting a two-legged gait would have been uncomfortable, inefficient, and dangerous for them. Yet, according to the evolutionary story, many generations of quadrupeds did this.

Natural selection works in the present, so it is a misnomer to attribute forethought to it, as is often the case. The urge to give agency to evolution and natural selection is also common, and also incorrect. I hate to personify natural selection at all, but I will do so here. Why should natural selection preserve adaptations of no immediate use to a group of primates?

The answer cannot be that the changes might have future utility. Yet, that is the inevitable conclusion. Over many generations, natural selection changed a group of primates to make them less fit for their environment? That is the opposite of how natural selection works! The optimized structures and systems for two-legged walking in humans need a good explanation.

That reason cannot be a single pressure. It also isn't good enough to combine several pressures into one hypothesis. One must show a *series* of circumstances to induce the sequence of changes for producing bipedal gait. No one subscribed to the evolutionary story attempts to do this.

An accepted evolutionary explanation for ancient primates adopting two-legged walking may be impossible. And even if one emerged, it would not be testable. But, let's say that there is a sufficient explanation. In that case, it would be helpful to explore how the human body and nervous system could have co-evolved. The development of effective two-legged walking requires the co-evolution of hardware and software.

Co-evolution of Optimized Features is Implausible

The evolutionary story is that humans and chimpanzees come from a single ancestor. This ancestor

46

was allegedly a primate that walked on all fours. In chapter 2, we saw that humans have several features for walking on two legs that are unlike any other primate. It would be remarkable if gradual changes over time produced just one of those features. For humans to evolve bipedal gait, we need much more than that.

The co-evolution of new skeletal and muscular system features is needed. And, the nervous system must be modified to use the structural changes. In robotics, designers adjust the control system when there are hardware changes. It is also useful to change the robot's hardware to make controllers more effective. Co-evolving optimal features for two-legged walking means *simultaneous* body and neurological changes.

Before accepting this idea, there must be a mechanism to achieve it. Neo-Darwinian macroevolution is the most popular hypothesis for large-scale evolutionary changes over time. What is its mechanism for the changes? The natural selection of favorable traits produced by genetic mutations. This explanation has many flaws, but I will mention three.

First, rather than being an engine of useful change, mutations are harmful. Second, rather than creating new genetic information, mutations corrupt or delete genetic code. Third, mutations do not affect epigenetics, which is necessary for changing body plans. Macroevolution could not have done what its advocates say it did. Let's examine these more closely.

Mutations are harmful

The evolutionary story states that mutations are helpful to populations in nature. This does not align with the facts from a probabilistic perspective. Genetic mutations are generally negative or neutral. Useful mutations are rare.

For example, harmful mutations have accumulated in the human genome over time. Humans today have decreased cognitive abilities compared to humans of two to six thousand years ago[19]. Scientists observe a slow but inevitable decline in our physi-

---

[19]Gerald R Crabtree, *Our fragile intellect. Part I*, 2013.

cal and mental abilities[20]. The evidence shows that mutations lead to unfavorable changes over time.

I saw this in some of my previous research work. We used an optimization method called evolutionary algorithms. To get this method to work, we always used mutation rates that were much higher than what is in nature. We also had to use high rates of useful mutations, compared to what we observe in biology.

This really made me think. Using natural mutation rates made the method useless or hopeless to converge. What if the macroevolution (upon which the method is supposedly modeled) has never actually occurred in nature at all? I can say, from experience, that no macroevolution ever happened with those algorithms. Small changes to a data structure never changed it into anything other than a data structure with different information.

A quadruped would need to undergo major body plan changes to evolve into a biped. These kinds of

---

[20]Michael Lynch, *Mutation and human exceptionalism: our future genetic load*, 2016.

changes require mutations during early stages of development. Mutations that come later in development are too late to produce evolutionary adaptations[2122]. One developmental biologist saw a problem with embryonic mutations for major adaptations.

He wrote that "it has long been a major tenet of evolution (and development) that any change to early developmental stages would be extremely disruptive or lethal"[23]. The dilemma is that late-acting mutations would have no effect on creating new body plans. But, early-acting mutations kill the embryo. Clearly, mutations cannot be the mechanism for changing body plans.

Mutations delete genetic information

To develop new structures and systems, one needs large amounts of genetic information. But, there are a tiny number of genetic sequences that can

[21]Keith S Thomson, *Macroevolution: the morphological problem*, 1992.

[22]Stephen C Meyer, <u>Darwin's Doubt: The Explosive Origin of Animal Life and the Case for Intelligent Design</u>, 2013.

[23]Keith S Thomson, *Macroevolution: the morphological problem*, 1992.

produce useful genes and proteins. And, the set of potential combinations is vast. So, it is unlikely that mutations can be the source of that new genetic information. And, random mutations are not an efficient way to generate useful genetic sequences. The evolutionary response to this?

"Unlikely doesn't mean that it is impossible. Given enough time, random mutations could possibly produce functional combinations." That seemed fair enough. Until someone actually estimated the odds of randomly generating a functional sequence of amino acids.

One researcher estimated the ratio of functional proteins to potential sequences of 150 amino acids to be 1 in 10 to the 77th power[24]. This suggests that the probability of random mutations producing just one new, functional protein from all of the potential sequences of 150 amino acids is "1 chance in 10^{77}– that is, one chance in one hundred thousand, trillion, trillion, trillion, trillion, trillion, trillion."[25]

---

[24]Douglas D Axe, *Estimating the prevalence of protein sequences adopting functional enzyme folds*, 2004.

[25]Stephen C Meyer, <u>Darwin's Doubt: The Explosive Origin of Animal Life and the Case for Intelligent Design</u>, 2013.

One could say that random mutations might have produced a functional sequence of amino acids a few times over millions of years. Yet, the neo-Darwinian claim is that this must have happened *regularly* over the course of history. And that's just one impossible claim. I could show others, but it isn't necessary.

Besides, this raises an important point. When we wade into the details of evolutionary claims, these are the kinds of anti-scientific conclusions that often emerge. This is why people ask whether you *believe* in evolution. It's because it is a faith system. And to believe in it requires that you ignore its anti-scientific nature.

Mutations do not affect epigenetics

Epigenetics[26] is a term that refers to non-genetic factors that determine how genes are (or are not) expressed. Such factors can be internal or external, and later generations can inherit them.

---

[26]The Greek prefix epi means "above" or "beyond", so the term "epigenetics" refers to biological factors that are not genetic.

Epigenetics helps explain how identical twins can grow up with disparate health outcomes. One twin may have heart disease, because of the expression of a particular gene. The same gene may be dormant in the other twin, who never experiences cardiac issues. Although the DNA is identical, epigenetic factors affect gene expression.

Epigenetics research shows that much more information than what is in DNA is crucial in the development of body plans[27]. We have already seen that mutations are harmful and delete genetic information. But let's say, for the sake of argument, that they could generate new and helpful genetic information. This is only one component of what one needs to create new transitional forms.

The neo-Darwinian model doesn't account for the epigenetic information needed to develop new body plans. Early embryonic mutations don't produce favorable adaptations. But if they could, corresponding changes in epigenetic information would be needed. Neo-Darwinism doesn't address this at all.

---

[27]Stephen C Meyer, Darwin's Doubt: The Explosive Origin of Animal Life and the Case for Intelligent Design, 2013

In this section, we have found that random genetic mutations are almost always harmful. They corrupt and delete, rather than produce, genetic information. They do not affect epigenetic information, a crucial component in the development of body plans. The mechanism of mutations with natural selection for producing large changes over time fails as a hypothesis.

Small changes in an ancient primate over a long period of time do not explain the large differences between modern apes and humans. This is especially true when the explanation does not offer a sufficient series of selective pressures to drive the changes. Then, we find no evidence of these changes in the fossil record.

It is far easier to fall down than it is to walk effectively on two legs. So, how do we do it so well? The evolutionary story goes something like this: *At first, all primates moved on four limbs. Over a long period of non-deliberate, unplanned trial and error, some of them began to change. Mother Nature (via evolution) happened upon the synchrony of body and neurological features found in modern humans.*

But how? Why?

Don't worry about it. That is how the effective bipedal gait in humans came to be.

In this section, we looked into the details of that story.

1. Evolutionists admit that you will find no common ancestor of humans and chimpanzees in the fossil record.

2. They offer no plausible mechanism for optimized, co-evolved adaptations resulting in humans.

3. There is no attempt to provide a sufficient series of selective pressures to induce the development of those adaptations.

These are compelling grounds to discard the prevailing macroevolutionary explanation for human bipedalism. But, you may be wondering: what about the evolutionary progression from ape to human that is often shown, like in the image on the next page?

55

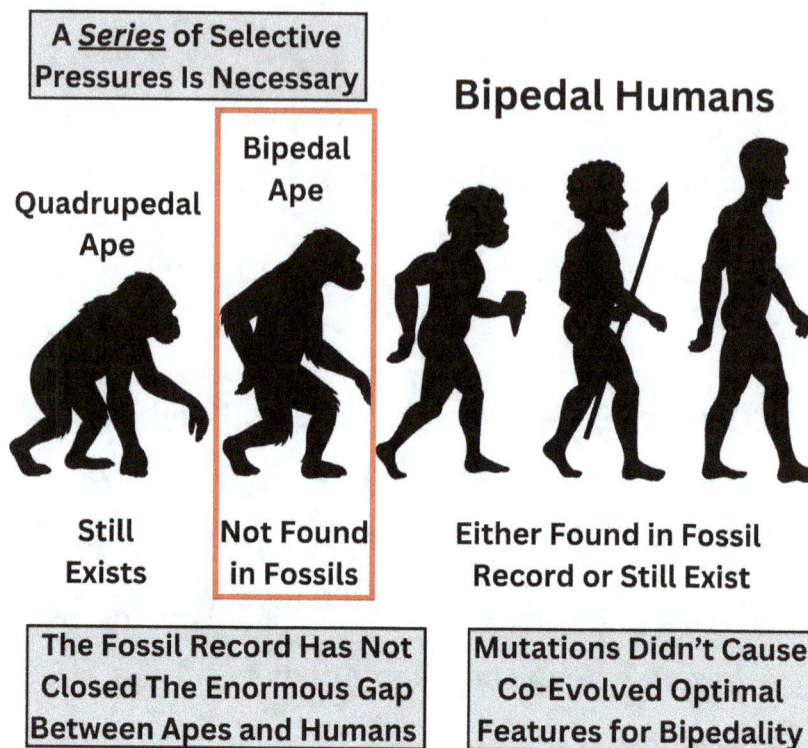

As far as I can tell, such images usually include creatures for which there is no fossil evidence. For example, a bipedal ape is often shown (like the one inside the red box in the figure), but no such creature exists in the fossil record. We see quadrupedal apes and we see bipedal humans in the fossil record, but we do not see bipedal apes.

But, for the sake of argument, let's imagine that this progression was actually shown in the fossil record. The evolutionary story would need a *series* of selective pressures to explain the dozens, if not hundreds, of modifications. Let's say we ignore the fact that no one offers an explanation like that. Then, we find that the claim of genetic mutations as the evolutionary mechanism is anti-scientific.

Do You Still Believe in Evolution?

The question of whether macroevolution has ever happened is for another book. But even if it did, it could not have produced the reliable bipedal balance and gait of humans. I find that there is a growing number of former evolutionists who recognize this problem. They are seeking a more plausible explanation than the neo-Darwinian synthesis.

In their conversation with Peter Robinson, mathematician David Berlinski, computer scientist David Gelernter, and philosopher of science Stephen Meyer agreed that Darwin's big idea is outdated[28]. It has become obvious that macroevolution is essentially an impossibility. Some, like Gelernter, once viewed Darwin's extrapolated evolution as a beauti-

---

[28]Mathematical Challenges to Darwin's Theory of Evolution [2019]: https://www.youtube.com/watch?v=noj4phMT9OE&t=720s

ful theory. But now, they can no longer justify their belief in it.

The idea that the small changes we observe in species have produced the biological complexity and diversity of the world is indefensible. One must ignore the scientific evidence to believe in an idea from a time when little was known about the cell or DNA. How about you?

Do you still believe in evolution?

Regarding Meyer's book <u>Darwin's Doubt</u>, Gelernter, who disagrees with Meyer and rejects the concept of intelligent design, felt compelled to say:

In this book, intelligent design is not a way to bring in a theological argument. It is a scientific approach, purely and absolutely valid scientifically. One can agree with it or disagree with it, but one doesn't have to reject it insofar as theology making an illegal move.[29]

[29]Mathematical Challenges to Darwin's Theory of Evolution [2019]: https://www.youtube.com/watch?v=noj4phMT9OE&t=1649s

Perhaps you believe that arguments for intelligent design are theological and unscientific. If so, then you have probably ignored them. You may disagree with my conclusions in this book. But I trust that like Gelernter, you will recognize the scientific evidence and arguments presented. You can deal with them fairly and without bias.

In this chapter, we have shown that neo-Darwinian evolution falls apart when examined closely. In the next chapter, we will show that a designing Intelligence is the most plausible explanation for our ability to balance and walk on two legs.

CHAPTER 3 SUMMARY

- Darwin was right in showing that natural selection causes changes in species, and that all current species were not independently created.

- Darwin was wrong in extrapolating small changes among species to the unfounded conclusion that all species are descended from a common ancestor.

- A lack of transitional forms is indefensible. Darwin believed that one day the fossils would turn up to provide evidence for his hypothesis. It's been over 150 years, and they still haven't turned up. If Evolution were true, the fossil record would be filled with missing links. Their absence can't be explained by the macroevolutionary hypothesis.

- Proposed selective pressures are insufficient. In fact, evolutionists make no attempt to provide a series of selective pressures, which would be necessary for producing the alleged series of adaptations between different kinds of lifeforms.

- Co-evolution of optimal anatomical and neurological features for efficient bipedalism is implausible. It takes a great deal of faith to believe that mutations (which are universally harmful) produced hundreds of beneficial changes in the human nervous system and anatomy that resulted in their amazing synchrony.

- Scientists are openly questioning and rejecting the neo-Darwinian model, and looking for a more plausible explanation for the complexity and diversity of life that we all observe.

Do you still believe in Evolution?

4 - Walking on Two Legs Has Been Designed

For they will regard this body as a machine which, having been made by the hands of God, incomparably better ordered and has within itself movements far more wondrous than any of those that can be invented by men.[30]

René Descartes

Theists are not the only ones who recognize apparent design in nature. The opening of the book <u>Why Evolution is True</u> states, "If anything is true about nature, it is that plants and animals seem intricately and almost perfectly designed for living their lives... Nature resembles a well-oiled machine, with every species an intricate cog or gear"[31]. Many scientists and philosophers conclude that this is evidence of design in nature.

Others argue that the appearance of design is an illusion. This latter idea relies on a materialistic ex-

---

[30]From part 5 of <u>Discourse on the Method for Conducting One's Reason Well and for Seeking Truth in the Sciences</u> (1637), translated by Donald A. Cress.

[31]Jerry A. Coyne, <u>Why Evolution Is True</u>, 2009.

planation. Macroevolution via natural selection and mutations is the most popular explanation. But, in the previous chapter, we saw three fatal difficulties with the evolutionary story.

This chapter presents an alternative explanation: the design inference. I also offer two aspects of human bipedalism that point to a Designer. Finally, I introduce the inverted design syllogism.

The Design Inference

The design inference is the idea that one can conclude that something has been designed from its characteristics. Natural causes don't produce characteristics like complexity, organization, and purpose. So, if one observes something with those attributes, it is safe to assume that it wasn't produced by nature.

Over the centuries, many philosophers and thinkers have proposed that nature shows evidence of design. Past advocates of the design inference include René Descartes, Johannes Kepler, and Robert Boyle. But, William Paley's 19th-century watch analogy is perhaps the most cited description of the design inference.

Paley showed that it might not initially seem absurd to assert that a stone encountered in a forest had always been there. But, such an assertion would not be reasonable to make in the case of a watch[32]. This would be true, even if one had never seen a watch before, and if parts of the watch were missing.

Paley further pointed out that a person encountering a watch for the first time "would be surprised to hear that the mechanism of the watch was no proof of contrivance, only a motive to induce the mind to think so"[33]. Imagine you found a pocket watch for the first time on the ground surrounded by grass, mushrooms, and ferns.

[32]Natural Theology: or Evidences of the Existence and Attributes of the Deity. Collected from the Appearances of Nature, 1829.

[33]Natural Theology: or Evidences of the Existence and Attributes of the Deity. Collected from the Appearances of Nature, 1829.

You would know (as Sesame Street taught us) that "one of these things is not like the other". In other words, the idea of "apparent design" would not seem reasonable, even if you were seeing a watch for the first time.

One form of logic used in the design inference is *a fortiori*, that is, an argument giving a stronger reason than one already accepted[34]. We can all agree that software, machines, and systems of systems have been designed. Arguing *a fortiori*, anything that is more complex, well-organized, and effective than those things must have also been designed.

- From biology, we observe that DNA far exceeds any software ever devised.

- Compared to any existing pump, the human heart is a marvel of mechanical engineering.

- And as we have observed so far in this book, the human body is an amazing system of systems. It is optimized for bipedal gait, and is more effective and efficient than any robotic walker.

Stephen Meyer pointed out that minds are the only sources observed to produce information-rich systems. Because of this, "one can detect (or, logically, retrodict) the past action of an intelligent cause

---

[34]Cambridge Dictionary: *a fortiori*. URL: https://dictionary.cambridge.org/us/dictionary/english/a-fortiori

from the presence of an information-intensive effect, even if the cause itself cannot be directly observed"[35].

For example, writers and computer programmers produce books and software. These products don't come into existence on their own. The existence of a book or a line of code is evidence of an author or a coder. Similarly, human bodies uniquely optimized for bipedal gait are what Meyer would call an information-intensive effect. Such effects are best explained by design.

The most common response to the design inference is dysteleological, a long word that simply means "badly designed". As an engineer, my responses to this line of reasoning can range anywhere from mild irritation to rage. For example, the ankle can be vulnerable to the weakness of the Achilles tendon.

Based on this, one author (a biology professor) asserted that "a modern mechanical engineer would

[35]Stephen C Meyer, *DNA by design: an inference to the best explanation for the origin of biological information*, 1998.

never design a joint with such an obvious liability"[36].

My response: *How could you possibly know that? You aren't a mechanical engineer!*

What "evidence" was presented to back up this assertion? The author observed that many athletes and sports participants hurt themselves by injuring Achilles tendons and anterior cruciate ligaments (ACLs).

My response: *Since it's such a bad design, why don't you come up with something better?*

One can imagine how critics might extend the dysteleological concept to the entire human body. In their estimation, a supremely intelligent Mind would never design a body with such weak knees, ankles, and backs. One book title claims that Evolution explains the existence of the human body and that Intelligent Design does not[37]. Predictably, the

[36]Nathan H Lents, Human Errors: A Panorama of Our Glitches, From Pointless Bones to Broken Genes, 2018.

[37]Abby Hafer, The Not-So-Intelligent Designer: Why Evolution Explains the Human Body and Intelligent Design Does Not, 2015.

author of that book (another biology professor) has no background or training in design or engineering.

Non-engineers like these often speak with confidence about how the human body should function and be organized. Never mind that they have no background in engineering design. I have not encountered this line of reasoning from an engineer. From our training, we understand that even a perfect design will have critics.

Paley addressed the "bad design" argument over two hundred years ago. In his analogy, he pointed out that an imperfect watch does not invalidate the fact that it had been designed. When non-engineers cite poor design in nature as evidence of evolution, they miss an important point. If poor design exists in nature, it is still *design*. Furthermore, in chapter 3, we showed that macroevolution isn't capable of design at all!

The graphic on page 71 shows the counterintuitive logic of the materialist.

- They admit that computer code has been written. But the far superior genetic code (DNA) came about through random processes.

- A water pump has obviously been invented. But the more efficient and effective human heart came about via mutations over time.

- Any humanoid robot is a deliberate result of engineering and design. Yet the more impressive human body must have evolved with no thought behind it.

Engineers, particularly those who design humanoid walking machines, understand that humans balance and walk on two legs unbelievably well. Humans do this, often in the midst of difficult terrain changes and unexpected disturbances. Our comparatively feeble engineering attempts to achieve bipedalism are clearly examples of design. Arguing *a fortiori*, it follows that the most capable bipedal walker in existence has also been designed.

Code? Written.

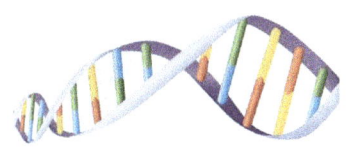

DNA? Random.

Pump? Invented.

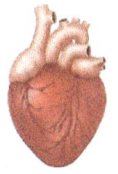

Heart? Mutated.

Humanoid Robot? Designed.

Human Body? Evolved.

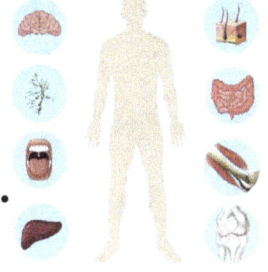

After chapter 3, we discarded the popular evolutionary story. It is most likely, to me and many others, that the human body's optimization for bipedal gait is the result of design. If you still aren't with us yet, in the next section, we'll show two aspects of bipedal balance that are evidence of a Mind.

Bipedal Balance Points to a Designer

In chapter 2, we saw that bipedal balance is *hard*.

In chapter 3, we found out that the evolutionary story does not explain humans standing and walking on two legs. In the last section, we observed design in nature. We also dealt with the "bad design" argument.

Here, we want to show two simple pieces of evidence for a Designer from the bipedal balance of humans. First, designers put redundancies in the systems they make. Second, learning processes must be designed, and cannot be randomly generated.

Designers Create Redundant Systems

The human body is a system of systems. In anatomy, you probably learned about the skeletal system, the respiratory system, the nervous system, and so on. All of these work together to produce an amazing bipedal balancing and walking machine.

Earlier, I mentioned that if I were to push you off-balance, you would quickly be able to regain balance. We have little understanding of exactly what you would do, but we understand how you sense the instability and try to correct it. There are a number of systems that work together that alert you to being off-balance. I will mention four of them.

1. You will feel the force applied to you at the place where you were pushed.

2. You will feel your weight shift at your feet.

3. Your vision (assuming your eyes are open) will alert you to a movement in your body that was unexpected.

4. Your vestibular system (in your inner ear) will give you a sense of a change in balance.[38]

Your brain processes all of this information and produces a reflex response (like swinging your arms, bending at the knees or hips, or taking a step) that allows you to regain your balance. This is a redundant system, meaning that if one of those sensing systems doesn't work properly, you can still compensate with the others. And, there are other

---

[38]National Institutes of Health, In brief: How does our sense of balance work? [2023]: https://www.ncbi.nlm.nih.gov/books/NBK279394/

redundant systems that cooperate to allow you to walk properly.

Otherwise, a tiny breakdown (like a sore foot or injured ankle) would prevent us from being able to walk at all. In engineering, we intentionally create redundancies in well-designed systems to account for smaller malfunctions and to avoid catastrophic failures. It is clear that the Maker of the human body did this also.

Learning is Designed, Not Random

In my research work, I designed controllers that changed little by little over time to make the simulated humanoid better at balancing and walking. These "learning" processes always begin with a "neural network" composed of random weights. The weights are tweaked and changed during the process of learning. After thousands of falls, the set of weights will now produce behaviors for more balanced standing and walking.

I ran hundreds of thousands of simulations, so I know a little something about machine learning.

There are two facts to keep in mind when it comes to learning.

1. Every neural network needs to be modified to achieve balance. Never, not even once, was there ever a neural network with randomly generated weights that needed no modification. No randomly generated neural network has ever made a humanoid stand and walk without the need for learning.

2. There is no such thing as a 100% effective learning process. If someone were to claim that, then we (in the machine learning community) would immediately be suspicious. But, that is exactly what happens in humans. The most remarkable thing (to me) about humans balancing and walking on two legs is that 100% of healthy, able-bodied people learn how to do it.

As in the case of machine learning, no one is born already able to walk. But, unlike machine learning, 100% of healthy humans learn to effectively overcome the dynamic instability of two-legged balance for standing and walking. The learning aspect of walking is, for me, the most impressive part of human bipedalism.

Effective learning cannot happen through random processes. To say that learning is a natural process would be a reversal of the Second Law of Thermodynamics. In our universe, things tend toward chaos and disorder. But, learning decreases entropy. We can approach this from an engineering design point of view or from a thermodynamics perspective, but we reach the same conclusion.

A perfectly effective learning process for something as difficult as two-legged balance makes zero sense apart from an intelligent Mind.

The Inverted Design Syllogism

The logic of the design inference can be presented as follows:

<u>Design Syllogism:</u>

- Major Premise: Every designed thing originated from a designer.

- Minor Premise: The human body is a designed thing.

- Conclusion: The human body originated from a Designer.

I will admit to assuming the major premise here. If you don't agree with it, then I invite you to show me an example of a designed thing with no designer. I know that this is an inductive argument[39], but it is awfully powerful. Inductive reasoning argues for a general conclusion from specific evidence. In our experience, we know of no designed things without a designer, so we can reason inductively that there are no designed things without designers.

You might concede that it is still possible to have a designed thing without a designer, but you have as much evidence of that as you do for Susannah's Unicornland. Show me a designed thing with no designer, and I will happily discard this premise.

The minor premise is where we have spent the majority of our time. We have shown that balanced standing and walking on two legs is difficult, that Evolution could not have produced it, and that it must have been designed.

Since this proves beyond reasonable doubt that the human body is a designed thing, it follows that the

[39]Rahul Awati, What is an inductive argument?, TechTarget [2022]: https://www.techtarget.com/whatis/definition/inductive-argument

human body originated from a Designer. This is the positive argument for the existence of God as Designer of the human body.

Since we are dealing with a dilemma (either God exists or not), if we could disprove God's non-existence, then that would prove that God exists. In other words,

If the non-existence of God MUST be false, then the existence of God MUST be true!

Inverted Design Syllogism:

- Major Premise: No designed thing originated without a designer.

- Minor Premise: The human body is a designed thing.

- Conclusion: No human body originated without a Designer.

It isn't obvious that the conclusion to our inverted design syllogism is the same as the conclusion from the earlier design syllogism, so let's show our work:

- Conclusion: No human body originated without a Designer.

- Equivalent: The human body originated not without a Designer.

- Final: The human body originated from a Designer.

Whether approaching from the positive or negative direction, the conclusion is unavoidable: the human body originated from a Designer.

In this chapter, we have argued that the most compelling explanation for the design of the bipedal human body is an intelligent Designer. This isn't a new idea, but it is one that is gaining momentum among modern scientists.

Stephen Meyer's book <u>Return of the God Hypothesis</u>[40] argues scientifically for the existence of God, using modern insights from the fields of biology, cosmology, and physics. The book <u>Your Designed</u>

---

[40]Stephen C Meyer, <u>Return of the God Hypothesis</u>, 2021: https://returnofthegodhypothesis.com/

Body[41] persuasively presents medical and engineering approaches to identifying design in the human body.

Similarly, we have shown that the science and logic behind bipedal balance and gait in humans points to a Designer. So, you might be wondering: who is this Designer? That is a great question, but there are some gaps we should fill first.

We haven't discussed definitions of God or of faith. A reasonable standard of evidence and proof of God's existence hasn't been established. After addressing those concerns in the next four chapters, we will return to the question of the Designer's identity in chapter 9.

CHAPTER 4 SUMMARY

- Everyone, even evolutionists and atheists, concedes that apparent design exists in nature.

- The design inference is an *a fortiori* argu-

---

[41]Steve Laufmann and Howard Glicksman, Your Designed Body, 2022: https://discovery.press/b/your-designed-body/

ment: since in our experience, design is evidence of a designer, the greater designs we see in nature are evidence of a greater Designer.

- Designers create redundant systems, and we observe plenty of redundancies in the human body, particularly related to bipedalism.

- Learning is a designed process, not a random one. Because 100% of healthy, able-bodied people learn to walk, that is evidence of a Designer of inestimable intelligence.

- The evidence of design we see in nature disproves the non-existence of a Designer. The inverted conclusion is equivalent to proof of the existence of the Designer, God.

Do you agree? What is your response to this proof of the existence of God?

5 - What is God?

Atheism is indeed the most daring of all dogmas, more daring than the vision of a palpable day of judgment. For it is the assertion of a universal negative; for a man to say that there is no God in the universe is like saying that there are no insects in any of the stars.[42]

G.K. Chesterton

Before showing how one can prove the existence of God, it would be a good idea to define what is meant by "God". Often, these kinds of discussions degenerate quickly into arguments about definitions, and we don't want to do that. This chapter is entitled "WHAT is God?" not "WHO is God?".

Both are important questions, but it makes sense to define God before identifying God. In this book, we argue that the design of the human body can be traced to a Designer. So, Evolution is not the so-called blind watchmaker, as Richard Dawkins has famously asserted. Beginning here and proceeding

---

[42]Charles II, Twelve Types [1902].

forward, I will describe and define this Designer by observing and analyzing the Designer's work. Later, in chapter 9, the Designer's identity will be revealed.

Defining "God"

First, God is an immaterial being. In other words, the Designer doesn't have a body or physical form. This is evident, and is the primary reason for the entire controversy between the atheist and the believer. The atheist believes that one cannot reasonably trust that an invisible Being exists without a physical form to inspect or observe.

Second, God is responsible for the existence of the universe, that is, God is its Creator and Designer. There are two other popular theories about the origin of the universe. I should address them now, so that they don't arise later in this discussion. Skeptics have long proposed that the universe might either be eternal or self-existent.

This creates a neat trilemma regarding the existence of the universe.

1. The universe could be eternal (that is, it has always existed and will always exist).

2. It might be self-existent (in other words, at one point it didn't exist, then it brought itself into being).

3. It may have been created (specifically, it had a definite beginning brought about by some external action).

There are no other options.

The eternal universe hypothesis directly opposes the Second Law of Thermodynamics. The Second Law contains the idea of entropy, and explains that matter tends toward states of lesser order and usable energy. If the universe is becoming more chaotic and contains increasingly less usable energy, it will eventually cease to exist.

The Cosmological Trilemma

Multiverse hypotheses have become popular, but they only kick the can down the road. There are really only three options for explaining the origin of the universe: it is either **Eternal**, **Self-Existent**, or **Created**.

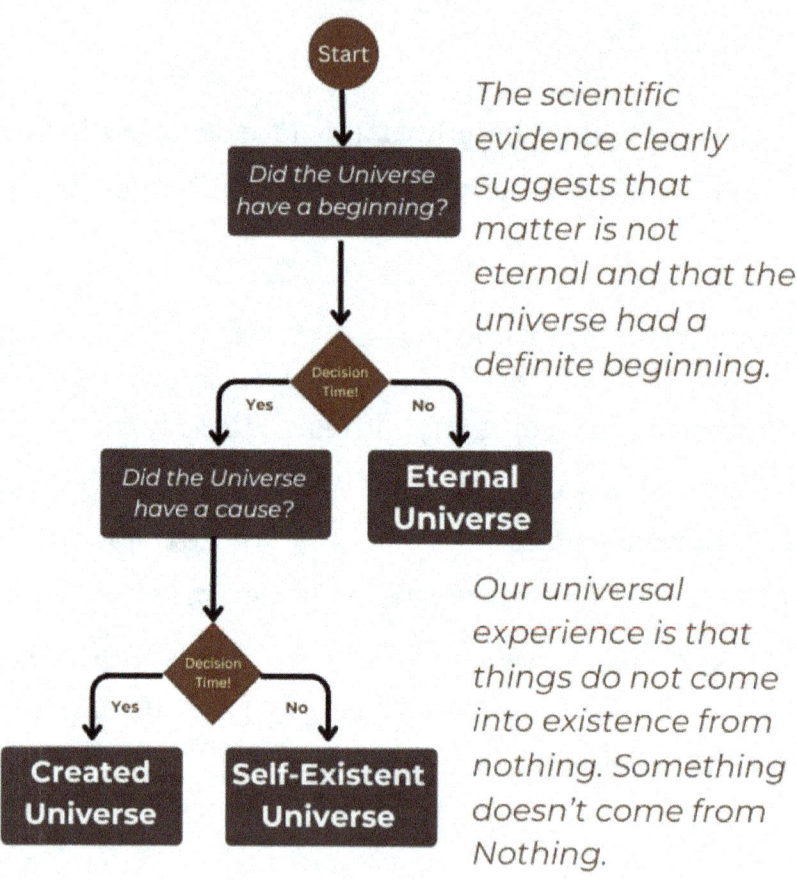

The scientific evidence clearly suggests that matter is not eternal and that the universe had a definite beginning.

Our universal experience is that things do not come into existence from nothing. Something doesn't come from Nothing.

But, if it is eternal and has always existed, this process of increasing entropy should have already met its end. But, it has not. The eternal universe concept has fallen out of favor among skeptics. This makes sense, because it is so diametrically opposed to what we observe and experience. Believers in materialism prefer to trust in observable and experiential evidence, so an eternal universe can be eliminated as an option.

That leaves us with a dilemma: either the universe brought itself into being or an external agent caused its beginning.

I have found that the idea of the universe creating itself to be conceivable only among those initially opposed to the existence of God. Where does one encounter anything that creates itself? Well, we don't see it. Ever. Clearly, it takes a great deal of faith to put one's trust in such an incredible idea.

In a 2020 interview[43], William Lane Craig commended the method of Alex O'Connor (also known as CosmicSkeptic), who, with great difficulty, attempted to make a consistent case for a self-existent universe in response to the *kalam* cosmological

---

[43]William Lane Craig and CosmicSkeptic Discuss The Kalam Cosmological Argument [2020]: https://www.youtube.com/watch?v=eOfVBqGPwi0&t=4080s

argument. An inductive argument against the self-existent universe is actually very simple and powerful. We never see self-existent things in this universe, so why should one think that the universe is self-existent?

O'Connor pointed out that even if there are no cases of self-existent things within our universe, it might be possible that the universe itself has a property that allows it to be self-existent. Craig then exclaimed that the intellectual price tag of such an idea is much too high for him to consider it. As the title of the book by Geisler and Turek suggests, Craig doesn't have enough faith to be an atheist[44].

Is there something missing in our definition? You might have expected to see attributes such as goodness, timelessness, and comeliness in our definition of God, but none of these are necessary.

I repeat: it is not necessary to include morality, immortality, or beauty in our definition of God. I am

---

[44] I Don't Have Enough Faith to Be an Atheist, Norman Geisler and Frank Turek, 2004.

not saying that God does not have any of these attributes. I am saying that it is not apparent from our observation of the created world that its Creator must be morally upright, immortal, or beautiful.

Here, we are attempting to draw unavoidable conclusions about the Creator from the Creator's designs. While we can safely justify ascribing immense power and intelligence to the Creator, it would be a stretch to require morality or immortality. We wish to be rigorous in what we assume and establish throughout this discussion.

What Does the Universe Have to Do With Designing the Human Body?

At this point, my atheist friends might raise two valid concerns. The first concern is that it isn't obvious that a universe that was externally caused is related to the apparent design of the human body.

Furthermore, I might be accused of begging the question by preparing to present God as Designer of the human body, after first assuming that God created the universe. How do we know that the same Agent who caused the universe to come into existence is the same One who created the human body?

First, the design of the universe and the design of the human body ARE related. The anthropic principle in cosmology asserts that the universe is uniquely fitted for human life. This book is about the apparent design of the human body. So, it is absolutely relevant that an Agent might have designed the environment in which we humans find ourselves walking around on two legs.

Second, there are some people who don't find this argument (that an external Agent must have created the universe) to be persuasive. The design of the human body is stronger evidence for many.

Even if you don't believe that the universe was created, I am counting on you being unable to ignore the design of the human body. But, if you already agree that there must have been an external Agent that brought the universe into existence, then you must also concede that the Agent (God) must exist!

A second concern might be that our definition of God allows for the alien or extraterrestrial being. My answer to this is: of course, it does! An alien is simply a being that is not a product of this universe. To characterize God as a (potential) alien is not a problem at this point, because we are simply establishing definitions. Identifying the Creator will come later. But, we should absolutely ignore any preconceived notions of aliens (for example, from the X-Files).

Also, we will generally refer to God as singular and masculine (even before offering a formal identification). But, this definition does not superimpose my admittedly monotheistic and patriarchal sensibilities. In short, our definition of God allows for the possibility of multiple beings of non-masculine gender. Yes, our basic definition allows for God to potentially be a team of She's or It's.

So again, we define God as an immaterial being that is responsible for the existence of the universe.

Here, we recall Chesterton's words that opened this chapter, which highlighted the bias and audacity of some atheists. The question of the origin of the universe is clearly a weak point for the materialist hypothesis in which the atheist trusts. Yet, there are some atheists who quite vociferously oppose non-materialistic explanations, like the existence of God.

How, exactly, does one who claims to trust only in what can be observed believe in the possibility of eternal matter or a self-existent universe? Neither of these can be observed, but both are claimed to be superior to or likelier than the idea of the existence of a Designer. But, why? Intellectually, an invisible God is as plausible (if not more so) as eternal matter or a universe that created itself.

Yet, some atheists can be fascinatingly creative (when imagining eternal matter or a self-existent universe). Then, they are incapable of conceiving of a theistic explanation for the cosmos that they could find plausible.

If God Exists

This is a good time to address the question of your response to God. Let's say that there is a God responsible for creating the universe, the Earth, and... you. What would your response to this realization be? How should you respond to this new knowledge? Have you thought about it? It's a good idea to think about this now, before we set out to determine what belief means, what reasonable evidence looks like, and how we intend to prove God's existence.

Let's say that I persuade you that you have been designed by God. Wouldn't you want to know more?

I am a robotics engineer. Most of my work has been with designing balance controllers for bipedal robots, and now, I am more interested in creating and improving robot designs. It seems to me that if I made a sentient robot, it would want to know who I am and why I made it. That seems to be a great place to start for you, as well.

I assume that you want to know the God who made you, and why He made you. Questions of identity and purpose aren't unique to believers. Everyone wants to know: Who am I? Why are we here? Where did we come from? Materialism and atheism have rather uncompelling answers to these questions: you are a randomly compiled blob of cells with no purpose that ultimately came from lifeless matter.

You're reading this book because you suspect that that answer is either untrue or at least missing something. And you're right. We see purpose, order, and complexity that cannot be explained apart from a Mind.

That there is a God-shaped hole in every heart is often attributed to Pascal:

This he tries in vain to fill with everything around him, seeking in things that are not there the help he cannot find in those that are, though none can help, since this infinite abyss can be filled only with an infinite and immutable object; in other words by God himself.[45]

---

[45]From section VII, fragment 425 of <u>Pensées</u> [1670].

Many of us find that there is something within us that is lacking, so we seek something to the void. Pascal referred to that thing as God. There are obviously destructive fillers for that hole, such as drugs and alcohol, power grabs, lust, and greed.

There are also less obviously destructive things such as family, work, physical fitness, and philanthropy. I say that they are *less* obviously destructive, because it is still ultimately destructive to place anything other than God in that hole.

- You have probably encountered a young woman who lives to please her boyfriend, only to be dumped by him in favor of some other girl.

- Likely, you know a workaholic, who is excellent at his job, but is ruining his health.

- Then, there's the guy who obsesses over what to eat, how often to exercise, and optimal supplements to take, but hasn't paid enough attention to his relationship with his daughter.

- And, how often do we see philanthropists who spend generous amounts of time and money to help poor people, only to find that those people are worse off than they were before?

The point here is that all of these can be good things, but they are not the highest Good. They are not God. Whenever humans worship things or people rather than the Transcendent One, it is destructive. The God-shaped hole one seeks to fill with alcohol, good deeds, or greed will not be filled with any of those things. That's because it's a God-shaped hole. Only God can fill it.

So, perhaps your response to realizing the existence of God is that you will start seeking answers to some of the biggest questions in life. But, we're not there yet. We have spent some time thinking about what God is (the Designer of the universe) and how you might respond to an understanding that God exists.

In the next chapter, I will define one of the most controversial terms in any discussion or debate on the existence of God: faith. We will examine the nature of belief, and ask the question: what does it mean to believe in God?

CHAPTER 5 SUMMARY

- God is an immaterial being that is responsible for the existence of the universe.

- The scientific evidence suggests that the universe is not eternal, and in our experience, things do not pop into existence from nothing.

- The most reasonable explanation for the universe is that an external agent brought it into existence.

- The idea that God created the universe has fewer problems than theories of eternal or self-created universes.

What would be your response to finding out that a God made the universe and you?

6 - What is Faith?

He doesn't have faith. He has confidence. Confi-dence. Con-fi-de, from the Latin, "with faith".[46]

Sye Ten Bruggencate

After establishing what is meant by "God," it is necessary to define what it would mean to "believe" or "have faith" in God. Dictionary definitions of "faith" will include synonyms like belief and confidence, but perhaps the most common non-religious word that can serve as a synonym for faith is *trust*.

It is important to point out here that the believer's trust is not mystical. Often, atheists and skeptics will define the "faith" (of a believer) as an unin-formed trust in the untested or unknown, while the "confidence" (of non-believers) is said to be an in-formed belief in what is tested and known. While there are believers who speak of and live by so-called "blind faith", that is not what I mean here.

Confidence, faith, trust, and belief are synonyms.

---

[46]Epic Debate Over God's Existence [2015]: https://www.youtube.com/watch?v=aUKIVV48LOk&t=4680s

We might disagree on what one should have confidence in. We might not see eye to eye on how or why one should believe in a particular idea or person. But all of our lives are built on trust. Everyone has faith in something. If you can agree with that statement, then we agree on what faith is. If you don't agree yet, don't worry: I'll prove it to you.

Everyone Has Faith

Trust is a part of everyone's daily life. A person who boards an airplane trusts the plane and its pilots. He also trusts the air traffic controllers, flight attendants, and baggage handlers (well, maybe not them). He assumes that they will operate as intended, to transport him safely.

When a driver steps into her vehicle, she trusts that the vehicle will start, run, and transport her to the desired destination. While on her journey, she trusts that the vehicle in the adjacent lane will stay in that lane and not run her off the road. She trusts that the navigation app will accurately guide her to the place she wants to go. She believes this, *even if she has never been there before!*

The bottom line is that we all trust, believe, and have faith in many things that we interact with on a daily, if not hourly or minute-by-minute, basis.

I used AI to generate the above image (as well as the Fence image in chapter 1, and several other images you will see throughout the book). The woman smiles broadly, trusting that her car works properly and that the GPS is guiding her in the right direction. She clearly isn't worried about that oncoming tractor trailer in the lane beside her. But, I also know what you're thinking, and I agree: she really should keep her eyes on the road.

We could not function in this modern world, if we had to constantly evaluate every circumstance that required trust. No sitting in a chair you have never seen before. No eating food that you did not personally plant, grow, gather, or prepare.

You get the point.

Ok, so what? you might say. Who cares about those things? Most of those things that we trust in, on a daily basis, are largely trivial. But, we don't just trust in trivial things. There are more important things and people, in which and in whom we trust. What about your parents' love? How about your spouse's fidelity? Consider your own logic and sanity.

If we apply a commonly-used standard for faith in God, then there is no reason to trust that your parents care about you, that your spouse is faithful to you, or that your mind is not playing tricks on you!

Trusting Logic

Many atheists claim to be able to reason their way to a set of morals and ethics that the world generally agrees upon. Believers often point out that a consistent atheist should be amoral, and I would agree with this. However, the issue is that many Christians mistake amorality (having no moral standard) with immorality (having bad or corrupt morals).

I find many atheists to be highly sensitive to the charge of immorality (or even amorality). They are justifiably suspicious of religion in general because of the immorality of many religious people.

But, I repeat, why should the atheist trust in her own reason when it comes to the subject of morality? Upon what basis do many atheists judge the God of the Old Testament? Furthermore, why should anyone else trust in the atheist's reasoning and justifications?

The point here is that there are beliefs upon which you trust and base your life, for which you have little to no material justification. You might offer a practical explanation: "How could the world go on if everyone questioned every little thing and needed justification for everything?" Good point. But that does not negate your willingness to trust in many things without justification.

And I could as easily point to the practical utility of faith in a Higher Power. This argument is often used in the sphere of morals that we have been discussing here. Many would argue that faith in a Judge to Whom we must answer is a powerful deterrent to some of the selfishness and criminality to which we may be practically prone.

The iron rule ("Might makes right") is awfully appealing when you are strong and have no one more powerful to whom you must answer. But that is a side conversation.

I repeat, in a different way, why do you trust your mind? And why should we trust your logic? If we are simply the products of mindless, random constructions of matter, then life and intelligence (and consequently, logic) are meaningless.

But if there is a God who constructed the universe in an orderly way, our bodies in an effective design, and our minds to operate reliably, then it follows that we can reason together to come to conclusions that we otherwise would not have. You can change my mind about things, and I can change yours. I want to suggest here that you borrow a great deal more from theism (like your views on the mind and morals) than you might care to admit. And that's okay.

But let's not say that you reasoned your way there, or that everyone would come to the same conclusions if they just thought it through like you have. What you believe about logic and reasoning existed before you, and is not a part of the material world.

A friend has pointed out that a materialist might argue that human logic is simply the development of a bias toward survival. This expresses a *purpose* behind the development of logic, but materialism has no room for immaterial things like purpose.

Besides, how and why would random collections of matter (that is, humans) develop a bias toward survival? The natural progression of our world is toward disorder and death, not survival.

You trust in things a lot less reliable or likely than God. We have already examined your faith in your own reason and intellect. Most people trust in their spouse's faithfulness or a parent's love. No one thinks twice about their belief in navigation apps, stable chairs, and safe restaurant food (although food poisoning is pretty common).

The existence of all of these things necessitates a Creator. If you can believe that these things exist, then it follows that their Creator exists. In short, it is far more likely that a Creator exists than that your fast food burger won't make you sick. Notice that I am not saying that you should put as little thought into whether a Creator exists as you do into whether your burger has been tampered with. It is exactly the opposite of that.

I am saying that although you spend little time thinking about relatively important things (like the safety of food that will enter your body), you should spend even more time thinking about the most important thing: the existence of your Creator.

We will end this chapter with a thought to lead us into our next chapter: what kind of evidence would make you reconsider your position on the existence of God? It is quite possible that you have spent very little time considering this. Well-known atheists who have conducted debates on the existence of God have recently been dumbfounded by this question, which suggests that they have not seriously considered it. But, we are about to consider it in the upcoming pages.

What is it? What is that thing (that perhaps you dread encountering) that would make you rethink your take on God's existence? If you don't have a good answer yet, that's fine, we'll help you come up with one in the next chapter.

CHAPTER 6 SUMMARY

- Faith, belief, trust, and confidence are synonyms.

- We all have different beliefs and reasons for faith, but everyone believes in something.

- Our lives are built on trusting in things that we don't constantly evaluate, like the love of family, the fidelity of a spouse, or your own sanity and reasoning.

What kind of evidence would make you reconsider your position on the existence of God?

7 - Does Faith in God Require Extraordinary Evidence?

Extraordinary claims require extraordinary evidence.[47]

Carl Sagan

I recently came across an announcement for a local meetup of "freethinkers". The flier included the quote from Carl Sagan that opens this chapter: "Extraordinary claims require extraordinary evidence."

But is that true? What constitutes reasonable evidence for the existence of God?

What Evidence Do You Need?

If you are agnostic or an atheist, please do not gloss over these questions. Perhaps you believe or have been told that theists are simple- and close-minded, and that we simply blindly believe in an invisible God because it's what we have been conditioned to think. But, I can't think of any believers who have never questioned or doubted their faith in God.

---

[47]Broca's Brain: The Romance of Science, 1979.

I am now asking you to question your assumptions regarding God. If God exists, what kind of evidence would you need, to trust in His existence? Although I have never come across a believer who never questioned their faith, it is quite common to find atheists who are quite sure of themselves and express little to no doubt that God does not exist.

You can find many instances of prominent atheists confirming that they cannot conceive of evidence that could convince them of the existence of God. You should be highly suspicious of anyone who says that about anything. Think about it: what kind of scientist says that they cannot think of evidence to disprove an idea they hold to? Apparently, atheistic ones.

This is why the narrative of the battle between faith and science is a false one. Modern science originated from believers in God who had a desire to better understand the orderly Cosmos in which we reside. Science is incoherent apart from a theistic worldview, so if Science has an opponent, it is atheism, not faith. In this chapter, we will lay out what I believe to be the kind of evidence that should be compelling for the doubter, the agnostic, and the atheist.

But for now, answer the question. What evidence would YOU need to shake your unbelief? In this chapter, I will include stories from former atheists. They wrestled with this question, and then came across the evidence that changed their minds.

You should not continue to read this book until you have answered this question. What would it take to change your mind? We want to establish this now, so that the goalpost is not moved later. Would evidence of design in nature be compelling? How about a universal sense of morality in humans? Perhaps it is the generation of an orderly Cosmos out of chaos? Write it down in the section below.

Evidence That Would Convince Me of God's Existence

_____

_____

_____

You will have an opportunity to change your mind by the end of the chapter. It is quite possible that your bar is unreasonably high (like needing to see God write out His name in the stars), but you will be able to establish a reasonable standard before we proceed to the next chapters. It is reasonable to apply the same logic and reasoning to establishing the existence of God that one uses to establish other things in which he or she trusts.

Extraordinary Claims Do NOT Require Extraordinary Evidence

In the previous chapter, we made the point that there are many material things in which we trust, based on minimal direct evidence.

- You don't screen the pilots who fly your plane or the grounds crew that maintains it.

- You sit in your chair, ignorant of its designer

or manufacturer. Furthermore, you have no idea whether or not it has been sabotaged in some way prior to your sitting on it.

- You follow that app's navigation directions, although you have never been to your destination or on any of the roads leading to it.

There are other things in which we trust, without constantly second-guessing or requiring more evidence.

- You believe in the love of your parents. You do not ask for more evidence of this.

- You trust in the fidelity of your spouse (hopefully!). This trust doesn't require tests or surveillance. In fact, requesting more evidence in these relationships is often viewed as a sign of one's own guilt or of some deep-seated insecurity. In most cases, more evidence is not actually warranted.

- You have faith in your own reason and sanity (at least, I think you do!). In general, constant doubt of one's own sanity is a sign of insanity.

You should be mistrustful of anyone who cannot conceive of reasonable evidence for the existence of God. Such a stance is obviously biased against theism. Besides, extraordinary things happen all the time!

This is the most obvious problem with Sagan's assertion that extraordinary claims REQUIRE extraordinary evidence. Events out of the ordinary (sometimes called "black swan" events) happen regularly. Yet, non-observers of those things believe them to be true, based on quite ordinary evidence. Ironically, the atheist's requirement of extraordinary evidence is suspended for extraordinary materialistic claims like eternal or self-existent universes.

Carl Sagan passed away in the 1990's. But, surely he would find that what happened in the world due to COVID in 2020 to be an extraordinary claim. If extraordinary claims require extraordinary evidence, then if he stepped out of a time machine into our current time, would he believe what he saw?

What evidence could we offer to make him believe in this thing called COVID? We could provide a reasonable amount of quite ordinary evidence. We could offer eyewitness testimony, videos, and documents. But, if he truly required extraordinary evidence to believe in the extraordinary claim of COVID changing the world, then he might step back into his time machine to return to the 1990's!

Earlier, I mentioned that Susannah explained to me the difference between unicorns and Pegasuses (or is Pegasi the plural form of Pegasus?). It should have been obvious, but, again, in case you were wondering, unicorns are objectively better than Pegasuses. But for a moment, let's pretend that I do not believe in Unicorns.

It would be unfair of me to ask Susannah for evidence of unicorns, without first establishing what would constitute reasonable evidence of their existence. For example, I might ask to see a video of Susannah riding a unicorn. Assuming she presents me with a legitimate, undoctored video that was not generated by artificial intelligence, I should accept her evidence.

But, again, I should have some sense of the kind of evidence that would induce me to believe in unicorns, before rejecting their possibility outright. That is, if I didn't believe in them.

If I were to tell you that there is an invisible force that pulls two objects toward each other, it would not be obvious that such a thing actually exists. Yet most people believe that it does. We call it *gravity*. But, what is gravity exactly? We have not yet been able to observe this force, but we can see its effects, which is why we can speak with certainty about it without seeing it.

The existence of God is very similar to this. Believers admit that our faith is in an invisible Being, but we also point out that His effects are not invisible. Many former atheists have become persuaded of the existence of God after observing the evidence.

Two Former Atheists

Dr. Francis Collins grew up believing that faith and science were at odds. This led him, without much thought, to becoming agnostic, and eventually, an atheist. His unbelief was not because of a lack of evidence, but, rather, he required no evidence for his unbelief. An interesting thing happened as his understanding of science deepened:

I slowly and rather reluctantly came to the conclusion that belief in God, while not possible to prove, was the most rational choice available. Further-

more, I saw in the very science that I so loved
something that I had missed — the evidence that
seemed to cry out for a Creator: there is something
instead of nothing, the universe had a beginning, it
follows elegant mathematical laws, and those laws
include a half dozen constants that have to have
the exact value they do or there would be no possi-
bility of anything interesting or complex in nature.
God must be an amazing physicist and mathemati-
cian.[48]

David Wood was an atheist as a young man, and has pointed to his atheism as the basis for his former megalomania and homicidal thoughts. He went to jail after attempting to kill his sleeping father with a hammer. While in jail, he began to make a list of more victims to brutally murder upon his release. David's cellmate Randy was a believer, and their conversations began to affect David. His atheistic thoughts began to shift:

First, what's called the Design Argument finally hit
me. I was looking at a wall, and how the bricks

[48]Dr. Francis Collins: From Atheism to Faith [2020]: https://www.youtube.-com/watch?v=N_VN4Risnb4&t=100s

were arranged, and I thought to myself, you know,
if someone told me that these bricks went into this
order by some process that didn't involve intelli-
gence, I'd smack 'em in the mouth, and yet I be-
lieve that life formed without intelligence, when the
most basic living cell is unimaginably more compli-
cated than some bricks stacked on a wall. Why did I
blindly accept the extraordinary claim that life
arose spontaneously from non-life without demand-
ing some very good evidence?[49]

David and Dr. Collins both found the Design Inference to be compelling. This happened, despite the
two men coming from disparate backgrounds. Perhaps, you also noticed that neither man wanted to
become a believer. Each got there by examining the
evidence and making rational conclusions.

Set the Standard Now

So, in this chapter, I want you to establish a standard of evidence. What will it take to convince you
of the existence of God? It may be that you have not

[49]How God Destroyed My Atheism (Christian Testimony): https://www.y-outube.com/watch?v=jb2ggj9mKM0&list=PLGZlwRmiLjm3p7xYJN9Z9-CoPA-MwTQm4n&t=1238s

given this much thought, but it is crucial to any serious treatment of the subject. How else will you be able to recognize reasonable evidence for faith, if you have not first defined what kind of evidence will be reasonable?

I want to be clear that there are many people who have come to faith in God without having first considered this question. When such people describe their journey to belief, their experiences range from it feeling as if a switch was flipped to a long, arduous path in which they fought against their ultimate conclusion at each step of the way.

In your case, however, let's imagine you have come to faith in God. What would it take to get you there? And, what will you do upon encountering this evidence? At this point, I am assuming intellectual honesty on your part. If you think this through carefully, then you should not have the undesirable feeling of being tricked or duped. Everyone is entering this discussion with eyes wide open.

However, if you establish what "reasonable evidence" looks like now, and then it is presented to you, I am trusting that you will not "move the goal-post", so to speak. This is a term taken from American football.

In that sport, competitors play on a field that is one hundred yards in length. After a team scores a touchdown by moving the ball across the goal line and entering the end zone, it has the opportunity to attempt to kick the ball through the upright for an additional point.

This upright, or goal posts, is a fixed structure behind each end zone. But what if, as the kicker prepared to kick the ball, someone moved the goalpost to the left, or to the right, or even backward several yards? This would certainly affect the accuracy of the kick, and would likely cause the kicker to miss the attempt.

Similarly, I would like you to set your expectations for reasonable evidence now. If you encounter that evidence and change the expectations, you will have moved the goalpost. Let's not do that. After you read this book, I hope you will let me know if that standard of evidence was met for you. I want to end this chapter by asking you to consider what else in life meets the standard of evidence you are placing on belief in God.

You might well ask, why should anything in life (which by definition would be less than God) require as much or more evidence than belief in God? That is a good question. My point here is not to say that you should not require a great deal of evidence, but simply that your standard should be reasonable.

I often encounter standards that atheists set that are either unattainable or purposely unreasonable, so that the conclusion of belief in God can safely be avoided, even if true. I want to warn you against that line of thinking, since it could cause you to miss out on the most fundamental truth, simply because it is unsavory to you right now.

CHAPTER 7 SUMMARY

- Extraordinary claims do NOT require extraordinary evidence.

- Often, atheists and materialists come to their beliefs (or unbelief in God) without examining any evidence at all.

- Many former atheists have found the Design Inference to be compelling evidence of God's existence.

Do you have a bias, either for or against faith in the existence of God?

8 - How Do You Prove God?

...by commencing with those objects that are simplest and easiest to know, in order to ascend little by little, as by degrees, to the knowledge of the most composite things...[50]

Renė Descartes

In my sophomore year of college, I fractured my right wrist while playing a spring intramural basketball game. I kept playing and we won the game, but that was the end of the season for me. I had to wear a cast for the next few months, and was unable to hold anything in my right hand, like a pencil. My temporary disability meant that I could withdraw from the Beginner Piano class I was taking, but it didn't exempt me from Advanced Geometry.

In case you were wondering, that class wasn't about learning the difference between a square and a circle. Most of the semester was about learning to derive increasingly difficult geometric proofs. We

---

[50]From part 2 of <u>Discourse on the Method for Conducting One's Reason Well and for Seeking Truth in the Sciences</u> (1637), translated by Donald A. Cress.

would begin with a set of information and a desired conclusion that needed to be proven. Step-by-step, we were tasked with showing how the initial set of conditions could logically lead to the final, desired conclusion.

Because I broke my right wrist, I had to use my left hand for many things I would normally do right-handed, like writing. I tried to use AI to generate the following depiction of me in college with a fractured right wrist doing Advanced Geometry homework with my left hand. But, the AI wouldn't produce a picture of a young man with a cast on his right forearm, writing with his left hand. Believe me, I tried many times.

Sometimes, the AI put casts on both arms and one time, it gave the guy three arms! Each time, the person is shown writing with his right hand. It appears we need to train our AI's with more images of left-handed people. Anyway, it took me far too long to figure out that I could just flip the picture in an image editor to get the right effect. But, I finally did.

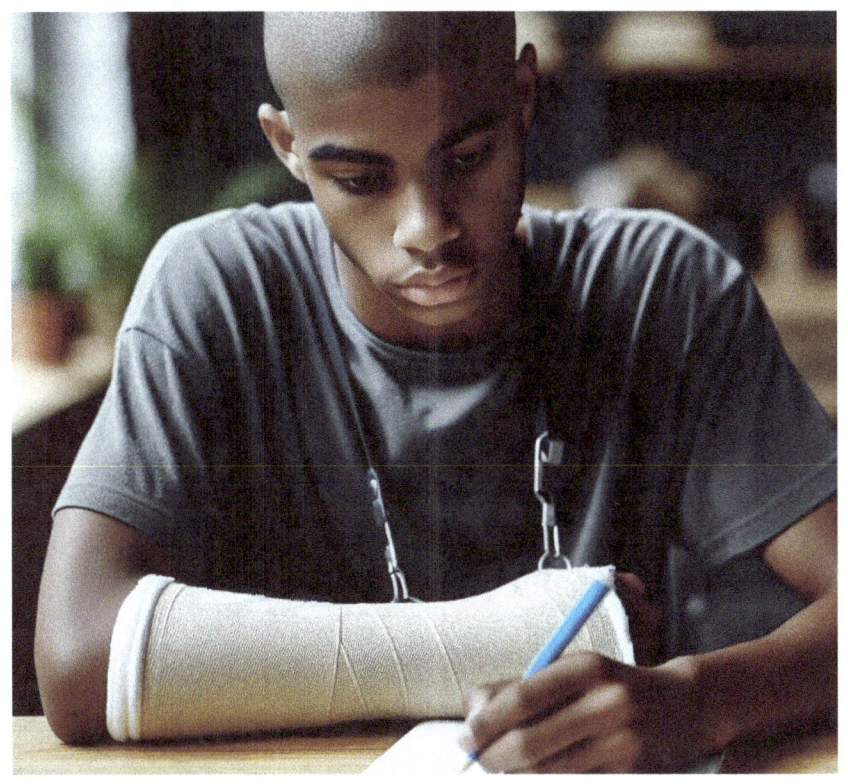

When doing my Advanced Geometry homework, I didn't have the luxury of scribbling down the wrong thing, because writing left-handed was slow. While deriving proofs, I learned to think several steps ahead. I would only write when I was sure that what I was about to write would take me in the direction of the final conclusion.

Otherwise, my homework would have taken several hours to finish each time. Over the course of my college career, I can only think of one or two grades I am as pleased with as my A in Advanced Geometry that semester. I had to get really good at methodically applying logic to prove mathematical conclusions.

Can we apply logic in this way to proving the existence of God?

Actually, yes, we can!

Descartes, the Apologist?

You probably don't read the writings of philosophers, especially not when they lived centuries ago. Yet, you are likely familiar with René Descartes. If you don't recognize his name, you may know of a phrase he wrote back in the 1600's:

I think, therefore I am.

Descartes realized that a person must first exist in order to think. Since he himself performed the act of thinking, then it must also be true that he existed. In the same discourse in which Descartes asserted this idea, he also offered arguments for the

126

existence of God. With all due respect to any Descartes fans out there, he seemed condescending toward atheists, and his reasoning for the existence of God was uncompelling to me.

But, he gave four useful laws outlining his approach to making logical conclusions. Here's how I would paraphrase Descartes' four logic laws:

1. *Avoid Bias:* Never accept anything as true that you don't know to be such.

2. *Divide and Conquer:* Divide each problem into as many parts as is required to better solve it.

3. *Simple to Complex:* Think in an orderly way, beginning with the simplest and easiest ideas, and taking baby steps toward more complex concepts.

4. *Documentation:* Number everything and then review, so as to not miss anything.

All these laws are useful, but for me, the third law is key, and the others are helpful in applying it. The quote that opens this chapter is Descartes' description of the third law. Descartes figured out what I learned in that Advanced Geometry course.

Making logical conclusions begins with simple, established concepts. Then, one proceeds step-by-step, in an orderly way, to a more complex conclusion. Following this procedure, the final result is just as true and established as the simple ideas at the beginning.

Proof Beyond Reasonable Doubt

We have already established a working definition of "God": an immaterial being that is responsible for the existence of the universe. We then described what we mean by "belief": to live in a way that exhibits trust in something or someone. These two definitions give us a sense of what is meant by the term "believing that God exists". It means to live in a way that shows trust in the immaterial Creator of the universe.

In the previous chapter, we set the standard of evidence for believing that God exists. Reasonable evidence should at least include an examination of the universe to determine whether a God is necessary to explain its existence.

The operative word here is "reasonable". In a court of law, a jury is tasked with trying a defendant for a crime, such as murder. In such cases, a conviction should only be handed down if the evidence establishes guilt beyond a "reasonable" doubt.

This is not always true for other violations of law or disputes between people. In those cases, a judge might make a ruling based on the evidence suggesting guilt more likely than innocence. For example, a defendant might lose a case on a 50.1% likelihood of guilt. But for serious crimes, such as capital offenses, the jury must not convict a defendant, if they are only 50.1% sure. That is *not* beyond a reasonable doubt.

Similarly, I would not ask you to trust in the existence of God, if you are 50.1% sure. You would retain a considerable, not insignificant level of doubt. Our proof of the existence of God must eradicate all *reasonable* doubt of His existence.

That is not to say that one will or can be necessarily 100% sure, as in a mathematical proof. But, one can be sure enough to no longer have reasonable doubt. Sure enough to live their life as if God truly exists, as Jordan Peterson has popularly asserted.

Bad Logic to Avoid

There are a few common logic pitfalls that should be avoided. The first is the appeal to authority. This often comes up in discussions like this. You quote your favorite atheistic scientist, and I appeal to my favorite theistic engineer. You one up me with this bright authority, and I one-up you with that believing thinker.

Of course, the problem here is that all six of us (you, me and the four authorities) could be wrong. Or, we might not be presenting clear, straightforward arguments. In a related vein, I want to point out, as have others, that capital-S Science does not "tell" or "teach" us anything. Personifying Science can be a way to assert what one's favorite scientists say. And if Science says it, then I cannot possibly offer a counterpoint.

This allows for many weak hypotheses to get passed off as theories and laws, when they are nowhere near being either. "Respectable" people might gasp in astonishment when you question the validity of an accepted hypothesis. For example, many scientists have accepted the hypothesis of all life forms descending from a common ancestor.

The same scientists have no problem with rejecting laws of science like the Law of Biogenesis and the Second Law of Thermodynamics. When evolutionists suggest hypotheses like abiogenesis (life beginning from non-life) or an eternal universe, that makes *me* gasp.

Another logical pitfall to avoid is going along with the majority. This is related to the appeal of authority, but the authority is the masses (perhaps, of scientists). I shouldn't have to say this, but I will anyway: truth is not established by a popularity contest or vote. This is clear in the area of morality. Murder and rape are universally viewed as wrong and evil. Yet, even if everyone agreed to allow murder and rape, that would still not make them acceptable.

We theists have a good reason for saying this. Atheists often say that reason brings them to know that these things are wrong or evil. I have not encountered a coherent argument as to how that can be the case, apart from God. But again, the moral argument is not within the scope of this book.

Free Your Mind

I completely understand that the burden of proof is on the theist. Atheists and agnostics are justified in wanting to see evidence or proof before leaving their unbelief. Similarly, those of you who do not believe in Unicornland will want to see some evidence of its existence before you are willing to discuss Susannah's adventures there.

To prove the existence of an invisible, immaterial Being, it is necessary to show that Being's direct effects on this world. This is crucial when discussing this subject with materialists, since they believe that matter is all there is. To persuade a materialist of the existence of the immaterial, one must show that the immaterial is the only plausible explanation for some material effect in this world.

So, yes, the burden of proof of theism is on the believer. I also want to highlight that it is unclear to me how the militant atheist argues for the non-existence of God. It is possible to prove that an idea or concept is untrue. But, to eliminate the possibility of God's existence is in itself an impossibility.

A finite, limited individual cannot know everything or be in every place, both of which would be necessary to prove the non-existence of a being. That includes Unicorns, for all you Unicorn skeptics! While you might not believe in the existence of Unicorns, you cannot actually prove their non-existence. The best you can do is to reject any evidence or proof that Susannah offers.

Using time and energy to argue against the existence of a being is baffling to me. But in the case of atheists, the potential existence of God threatens a deep-seated belief system. That's why so many care so deeply about this topic, in a way that you don't care about whether Unicornland is real.

I would like to speak to the religious atheist. Perhaps, so far, you have agreed to or conceded our definitions of God, belief, evidence, and proof. Another thing that we have also attempted to do is to open your mind to the possibility of God's existence.

It is probable that your mind has been closed due to various circumstances and influences. It repulses you to think that the God of the Old Testament is someone to whom you should submit yourself. I ask a variation of a question often posed by Frank Turek: if God were real, would you serve Him?

We'll revisit this question at the end of the book, but if your answer is no, I am asking you to rethink that. Your belief system is being challenged by the potential existence of God. If you are unwilling to potentially let go of that, then you are vulnerable. You will be willing to believe in anti-truths so that your belief system will remain intact. But, if what you believe isn't true, do you really want to live your life based on a lie?

Inverting the Design Inference

The late investor Charlie Munger was well known for his use of inversion in problem solving. As a young man, he became a weather forecaster for the Air Force. In his new job, he used inversion to determine how to best handle his assignment:

I said, How can I kill these pilots? Now that's not the question that most people would ask, but I want to know what the easiest way to kill them would be so I could avoid it. And so, I thought it through in reverse that way, and I finally figured... There are only two ways I'm gonna kill a pilot. So I'm gonna get him to icing his plane can't handle and that will kill him. Or, I'm gonna get him someplace we're gonna run out of gas before he can land, because all the airports are sucked in. And I just was fanatic about avoiding those two hazards. And if Kobe Bryant had had somebody like me, he'd still be with us.[51]

---

[51]Redlands Forum: Charlie Munger [2021]: https://www.youtube.com/watch?v=Rh1WCzfCP24&t=2718s

Munger didn't focus on how to read weather maps better or any number of other potentially useful actions. Instead, he figured out the worst actions he could take. He determined what he could do that would endanger the lives of his pilots. Then, he simply made sure not to do those things.

Munger often asserted that many problems are more easily solved by first inverting them and then figuring out the answer to the reverse of the original problem. The inverse of the solution to the reverse problem is an answer–often the best answer–to the original problem.

How might we apply Munger's Inversion Principle to proving the existence of God?

We could merely present evidence for the existence of God. But, we can also attempt to disprove the non-existence of God. What if it was possible to show that the non-existence of God (the atheist's position) is in conflict with basic ideas in science and philosophy? Since we are dealing with only two options, disproving God's non-existence is the same as proving that God exists. In other words,

If the non-existence of God MUST be false, then the existence of God MUST be true!

In this chapter, we have seen that the proof of the existence of God can include both positive and negative parts. Earlier, I argued for a powerful, intelligent Designer of the universe. I have also demonstrated that this Designer's non-existence cannot be true. This was done by showing that the idea of apparent design is nonsensical or anti-scientific. This is another form of the Design Inference. The logic can be put in the form of a syllogism:

- Major Premise: Every designed thing originates from a designer.

- Minor Premise: The human body is a designed thing.

- Conclusion: The human body originates from a Designer.

In a syllogism, the conclusion logically follows from the major and minor premises. To disprove the conclusion, one must show that at least one of those premises is false.

I have already shown that the belief that a designed thing exists without a designer (an inversion of the major premise) is false. Surely, by now, you will agree that it cannot be true that the human body is not a designed thing. Since the inversions are false, the original premises must be true, which leads us again to the conclusion:

The human body originates from a Designer.

In this book, we have driven home the design inference by highlighting a specific instance in nature. We experience and observe it regularly: humans walking on two legs. My thesis has been simple. *The existence of God is the most reasonable explanation for the incredible ability of humans to balance and walk on two legs.* We are now prepared to identify the most reasonable Candidate as Designer of the human body.

CHAPTER 8 SUMMARY

- Proving the existence of God requires an orderly, logical approach.

- Proving the existence of God means establishing the conclusion beyond reasonable doubt, rather than making a mathematical formulation.

- Proving the non-existence of God to be false is the same as proving God's existence.

If you were persuaded that God exists, would you seek to learn more, or reject Him?

9 - Yahweh is the Designer

For you formed my inward parts; you knitted me together in my mother's womb. I praise you, for I am fearfully and wonderfully made.[52]

Psalm 139:13-14

In chapter 5, we introduced the following definition of God: an immaterial being that is responsible for the existence of the universe.

We have since shown that this intelligent Designer is responsible for the human body, which is remarkably capable of balancing and walking on two legs. But is that all there is to know? Can we learn more about this Architect of bipedal walking machines by examining the human body? What is the Creator's Identity?

The Universe, Yahweh, or Somebody Else?

To simplify our approach to this question, I will pose it as a trilemma. This is often done to simplify options in various subjects, but especially in sports.

---

[52]Holy Bible: English Standard Version, 2001.

- Is the GOAT (Greatest Of All Time) in basketball Michael Jordan, LeBron James, or someone else?

- Do you think the defending Masters champion will win it again this year, or will it be Tiger Woods, or someone else?

- We might also ask about your favorite fruit juice flavor: is it apple, orange, or something else?

Such questions allow us to narrow many choices down to only three. We do this by taking two of the most popular options, and then lumping all others together into a single category.

The correct answer may be the catch-all category. Or, it may be one of the other two options. But, this approach simplifies such questions of identity. Now that we have established the existence of an intelligent Designer of the universe, we offer three options for the identity of the Creator:

1. The Universe is the Creator.

2. A god other than the God of the Bible is the Creator.

3. Yahweh, the God revealed in the Bible, is the Creator.

You might be saying to yourself: hey, I thought we eliminated the universe as a potential creator and designer of itself and of the human body. We did, back in chapter 5. But, that doesn't stop people from offering it as a viable option. Think about it: how often do you hear people attribute agency to the universe? One author has written a book called <u>The Universe Has Your Back</u>. That title suggests an idea similar to the Indian concept of karma.

The idea is that you should do good things, think beneficial thoughts, and have positive energy. If you do this, you can trust that the universe will produce favorable outcomes and results for you. While potentially a comforting thought, it doesn't negate that it's impossible for the universe to create itself.

I'll use another inductive argument. Show me self-existent matter and I'll allow the possibility of a universe that created itself. Otherwise, we can all agree that that didn't happen.

That narrows our three options to two:

1. ~~The Universe is the Creator.~~

2. A god other than the God of the Bible is the Creator.

3. Yahweh, the God revealed in the Bible, is the Creator.

Our dilemma: is the Creator the God of the Bible or some other god?

Comparing Yahweh to Other Gods

In a criminal investigation, law enforcement officials seek to establish means, motive, and opportunity to determine if a suspect is guilty of some crime.

"Means" refers to the ability to perform an action. If the cookie jar is now empty, we don't blame my seven-month-old niece. She doesn't have the means to take the cookies. Likewise, a god without immense power and intelligence would be incapable of creating the universe, and specifically, the human body with its bipedalism.

"Motive" has to do with the desire or reasoning to act in a given way. My family knows not to accuse me of eating the remaining cookies in the jar, because I am a carnivore (yes, that's right; I just eat meat). I may have the means to take the cookies (unlike my niece), but I don't want them, so I can be eliminated as a suspect.

In our case, a god that is impersonal and immoral has little to no reason to create humans like us, who have personhood and morality, and for whom good actions and thoughts are rewarded. This God must instead be personal and moral.

"Opportunity" deals with the time needed to take the given action. Let's say that my sister has been trying to rock the baby to sleep over the past two hours during which the cookies were taken. She might be able to take the cookies (means) and she enjoys eating cookies (motive), but she has had no time (opportunity) to commit the crime.

The God we are looking for must be timeless and immaterial. Otherwise, there would literally be no opportunity to create the amazing human body.

So, to state this in a different way, as we seek to identify the Creator, we are looking for Someone with the capability, characteristics, and confirmation to have designed the human body.

By capability, we refer to the means: powerful and intelligent. By characteristics, we mean the motivation: personal and moral. By confirmation, we are looking for opportunity: timeless and immaterial. We have organized these attributes of the Creator in the following chart.

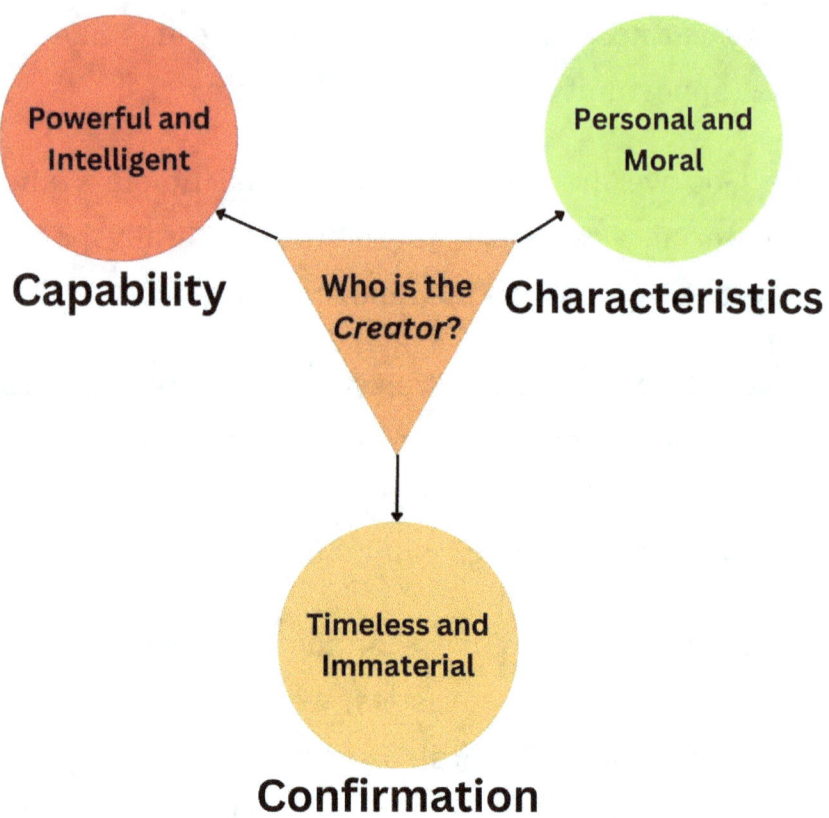

Confirmation

The Designer Must Be Yahweh

In this section, I am going to briefly show why the Creator of the universe and Designer of the human body must be Yahweh, the God of the Bible. The following passages are all from the World English Bible[53].

---

[53]World English Bible: https://worldenglish.bible/

<u>Capability</u>: These passages illustrate the power and intelligence of Yahweh.

Exodus 20:11–*...for in six days Yahweh made heaven and earth, the sea, and all that is in them, and rested the seventh day...*

Genesis 2:7–*Yahweh God formed man from the dust of the ground, and breathed into his nostrils the breath of life; and man became a living soul.*

<u>Characteristics</u>: These Scriptures speak to the personality and morality of Yahweh.

Psalm 139:1,4,13-16–*Yahweh, you have searched me, and you know me...For there is not a word on my tongue, but behold, Yahweh, you know it altogether...For you formed my inmost being. You knit me together in my mother's womb. I will give thanks to you, for I am fearfully and wonderfully made. Your works are wonderful. My soul knows that very well. My frame wasn't hidden from you, when I was made in secret, woven together in the depths of the earth. Your eyes saw my body. In your book they were all written, the days that were*

ordained for me, when as yet there were none of them.

Isaiah 40:10-11–*Behold, the Lord Yahweh will come as a mighty one, and his arm will rule for him. Behold, his reward is with him, and his recompense before him. He will feed his flock like a shepherd. He will gather the lambs in his arm, and carry them in his bosom. He will gently lead those who have their young.*

Confirmation: These texts describe the timeless and immaterial nature of Yahweh.

Isaiah 40:28–*Haven't you known? Haven't you heard? The everlasting God, Yahweh, the Creator of the ends of the earth, doesn't faint. He isn't weary. His understanding is unsearchable.*

Deuteronomy 4:15-16–*Be very careful, for you saw no kind of form on the day that Yahweh spoke to you in Horeb out of the middle of the fire, lest you corrupt yourselves, and make yourself a carved image in the form of any figure...*

Feel free to compare Him to whichever God you choose, but no other god has the capability, characteristics, and confirmation of Yahweh. So, we can safely identify Him as the Designer of the human body, and Creator of the universe.

In his first encounter with Yahweh, Moses had a similar question to the one we have been answering in this chapter:

Moses said to God, "Behold, when I come to the children of Israel, and tell them, 'The God of your fathers has sent me to you,' and they ask me, 'What is his name?' what should I tell them?" God said to Moses, "I AM WHO I AM," and he said, "You shall tell the children of Israel this: 'I AM has sent me to you.'" God said moreover to Moses, "You shall tell the children of Israel this, 'Yahweh, the God of your fathers, the God of Abraham, the God of Isaac, and the God of Jacob, has sent me to you.' This is my name forever, and this is my memorial to all generations.

Yahweh told Moses to refer to Him by another name: I AM. In this book, we have shown that as humans, we can take the evidence of the amazing human body that is able to walk on two legs, and reasonably conclude that it had a Designer. In this chapter, we have shown that the Designer is Yahweh, the great I AM. Our proof is complete, thus, the title of the book:

i walk therefore I AM.

By now, you should have come to believe that human bipedalism must have been created by Yahweh. But, we aren't quite finished. In the next and final chapter, we will end this book with a brief discussion of what one should do with such an important conclusion.

CHAPTER 9 SUMMARY

- The Designer must either be the universe, Yahweh, or a god other than the One in the Bible.

- The universe didn't design or create itself, so that leaves us with either Yahweh or a god outside of the One in the Bible.

- Whoever created the universe, and specifically the human body, must have the capability, characteristics, and confirmation to have accomplished this.

- Yahweh is revealed in the Bible as a Being who is powerful, intelligent, personal, moral, timeless, and immaterial. These are necessary attributes for the Designer we are trying to identify.

- The Designer of the human body and Creator of the universe must be Yahweh.

Do you agree? Can you think of a god that could possibly compare to Yahweh?

10 - So God Exists. Now What?

In God you come up against something which is in every respect immeasurably superior to yourself. Unless you know God as that–and, therefore, know yourself as nothing in comparison–you do not know God at all.[54]

C.S. Lewis

In chapter 1, I told you that I wrote this book for people standing at the Fence separating the colorful orchard of theism from the desert land of atheism. Perhaps you were at the Fence after wandering away from your faith. Maybe, you came to the Fence with curiosity about what is on the opposite side of atheism. In either case, if I have done my job, you are now walking away from the Fence, and into the orchard.

Welcome!

As we end this book, I want to show how far you've come, but also explain that the journey isn't over. C.S. Lewis spent several years making the journey

---

[54]C.S. Lewis, Mere Christianity, 1952.

from atheistic materialism to Christianity. As he expressed in the quote that opens this chapter, to know God is to understand Him to be Someone who is infinitely better than yourself in every conceivable way.

This book has highlighted God's creativity. This single aspect of His greatness is obvious when we observe bipedalism in humans. In chapter 9, we also saw that He is immensely powerful and unbelievably intelligent. He is closely personal and perfectly moral. God is unlimited with respect to time and space. Understanding these things about God means that you have already walked a long way from the Fence.

This changes everything.

But, the journey isn't over.

The land of theism is superior to the realm of atheism, because a fundamental reality is acknowledged here that is rejected there. But, there's even more truth to be discovered. Can I suggest your next book to read? Yes, you probably guessed that since I trust in the God of the Bible, that I will urge you to read the Bible. But, that's only partially right.

You may know that the Bible is a book composed of 66 books. The first 39 books are called the Old Testament, and the latter 27 books are known as the New Testament. But, which book should you start reading? Start at the very beginning with the book called Genesis, right? Well, you could do that, but it wouldn't be my recommendation.

Find the fourth book of the New Testament, written by John, one of Jesus Christ's disciples. Read that first. Then, go to the first book of the New Testament (written by Matthew), and read through the entire New Testament. You will come to the book of John again, and when you read it the second time, you'll find that it makes more sense.

After reading the New Testament, go to the beginning of the Bible, to the book of Genesis, and read through the entire Bible. When you have read the New Testament a second time (and the book of John for the third time), it will all make even more sense.

There are people in the land of theism that will tell you that all gods are the same, and that all of the religions teach the same thing.

That just isn't true.

When you read the Bible, you will find a God and teachings that are unlike anything you have encountered. Yes, there will be similarities, but the differences are so major that you just won't be able to say that Christianity is the same as Islam, Hinduism, Buddhism, or any of the other religions out there.

I would like to end with another quote from C.S. Lewis:

We may ignore, but we can nowhere evade, the presence of God. The world is crowded with Him. He walks everywhere incognito. And the incognito is not always hard to penetrate. The real labor is to remember, to attend. In fact, to come awake. Still more, to remain awake.[55]

In this book, I have asked you to see the presence of God in one of our most common and mundane tasks: balancing and walking on two legs. What you will find next, is that His fingerprints are everywhere. So, wake up, and recognize the presence of God!

---

[55]C.S. Lewis, Letters to Malcolm: Chiefly on Prayer, 1964.

Thank You!

Thanks for reading this book!

If you found this book helpful, I would greatly appreciate you leaving me a review on Amazon. This helps others find the book as well.

If you would like to contact me, my email address is jerry@jerrysweafford.com.

Last Update: September 27, 2024.